Strongly Interacting Particles

Chicago Lectures in Physics

strongly interacting particles

Riccardo Levi Setti and
Thomas Lasinski

The University of Chicago Press
Chicago and London

Chicago Lectures in Physics

Elementary Particles, by Riccardo Levi Setti (1963)
Group Theory and Its Physical Application, by L. M. Falicov (1966)
Experimental Superfluidity, by Russell J. Donnelly (1967)
Currents and Mesons, by J. J. Sakurai (1969)
Covalent Bonding in Crystals, Molecules, and Polymers, by James C. Phillips (1969)

The University of Chicago Press, Chicago 60637
The University of Chicago Press, Ltd., London

 Published 1973
Printed in the United States of America

International Standard Book Number: 0-226-47444-5 (clothbound)
Library of Congress Catalog Card Number: 73-83750

CONTENTS

PART II

Introduction to High Energy Phenomenology

A. Particle Exchange Processes

B. Connection between s- and t-Channel Phenomena

INTRODUCTION: ELEMENTARY PARTICLES IN PERSPECTIVE

A meaningful organization of an Elementary Particle Physics course of one quarter duration, given in 1971, posed problems of a very different nature from those encountered say eight or nine years earlier when this course was originally conceived by one of us (R.L.S.).
Examine briefly what happened over the years. By the mid-fifties (see, e.g., M. Gell-Mann and E.P. Rosenbaum, _Sci. Am._, July 1957) the world consisted of about 30 particles:

Strongly Interacting		Other	
Baryons + Antibaryons :	16	Leptons + Antileptons :	6
$(n, p, \Lambda, \Sigma^+, \Sigma^-, \Sigma^0, \Xi^-, \Xi^0)$		$(e^+, e^-, \nu, \bar{\nu}, \mu^+, \mu^-)$	
Mesons			
$(\pi^+, \pi^-, \pi^0, K^+, K^-, K^0, \bar{K}^0)$	7	Photon	1
total	23	total	7

By 1962-1963 (see, e.g., G. F. Chew, M. Gell-Mann and A. H. Rosenfeld, _Sci. Am._, February 1964) the accounting had already become considerably more complex, due to the proliferation of discoveries of resonant states of mesons and baryons. The strongly interacting particles, inclusion of particles and antiparticles, had reached a total in excess of 82, while the leptons and antileptons increased only by two to a total of nine (with the discovery of two kinds of neutrinos).

The Particle Data Group publishes yearly a Review of Particle Properties, in Rev. Mod. Physics and Physics Letters. Listings in 1970 (counting particles is not so easy anymore) showed approximately:

Strongly Interacting		Other	
Baryons + Antibaryons :	∿220	Leptons + Antileptons :	8
Mesons	∿ 60	(only a few, but very well known)	
total	∿280	Photon :	1
		total	9

Clearly there has been a phenomenal increase in the amount of experimental information available over the past ten years. This is true also in the field of weak and electromagnetic interactions, in spite of the non-proliferation of leptons! Correspondingly there have been exciting developments in the phenomenological (if not theoretical) description of particle systematics and properties. The successes of SU(3) and SU(6) have brought some and much needed order in the otherwise chaotic situation concerning the classification of strongly interacting particles. The successes of Regge phenomenology have correspondingly brought some order in the description of particle interactions at high energies, as well as having given another approach to particle classification. SU(3) made its way successfully also in weak interactions through Cabibbo's theory. In view of these developments, something drastic needs to be done today (which could not be done, say, in 1962) in order to avoid, in an eleven weeks course, either total confusion or total boredom.

The approach which was chosen necessarily transfers much of the burden to the reader, in the form of prerequisites, or parallel readings. Still, hard choices have to be made, to give a feeling for the state of the art, in face of at least a tenfold increase in the volume of relevant material.

The plan of the course is then to introduce the classification of elementary particles by means of the symmetry schemes, SU(2), SU(3), SU(6), quark model and related phenomenology. We then consider in detail some of the methods used to extract information on baryons from scattering experiments. The next topic will be Regge phenomenology and its connection to high energy physics on one side, and to particle classification on the other. Clearly the above plan is highly subjective and incomplete, reflecting only the inclinations of the instructor at the present time.

We wish to thank Nerissa Walton for preparing this manuscript for publication.

PART I. STRONGLY INTERACTING PARTICLES

A. CLASSIFICATION OF PARTICLES AND RESONANT STATES

1

GENERALITIES

Hadrons — Mesons: $\pi, K, \eta, \rho, \omega, \phi$, ...

Hadrons — Baryons: $N, \Delta, \Lambda, \Sigma, \Xi, \Omega$, ...

All mesons are Bosons, all Baryons are Fermions. The hadrons encompass the largest class of symmetries in the form of conserved quantities. Particles are correspondingly labelled by a set of quantum numbers. If we limit ourselves to consider particles (and anti-particles) with baryon number $B \leq 1$, the range of observed values for the good quantum numbers is:

Conserved Quantity	Symbol	Observed Values
Electric charge	Q	0, ±1, ±2 (times e)
Atomic mass or baryon number	B	0, ±1 (by choice)
Spin angular momentum	J	1/2, 3/2, 5/2, ... 0, 1, 2, 3, ...
Parity	P	-1, +1
Isospin	I	0, 1/2, 1, 3/2
Strangeness	S	-3, -2, -1, 0, +1 (for particles)
Hypercharge (Y = B + S)	Y	-2, -1, 0, +1 (for particles)
Charge conjugation	C	+1, -1 (for neutral systems with B = 0)
Isotopic parity	G	+1, -1 (for mesons with Y = 0)

The Gell-Mann Nishijima relation gives the particle charge in terms of I_3, B, S:

$$Q = I_3 + \frac{B+S}{2} \quad \text{or} \quad Q = I_3 + \frac{Y}{2} \tag{1}$$

where I_3 is the z-component of isospin I. For antiparticles $Y \to -Y$ ($S \to -S$, $B \to -B$).

Particles with $S \neq 0$ are called "strange." Strange mesons are the K-mesons, strange baryons, or <u>hyperons</u> are the $\Lambda, \Sigma, \Xi, \Omega$. The <u>nomenclature</u> which is adopted today already reflects the onset of some order in the multitude of particle states. In fact particles appear only in a limited number of <u>isospin multiplets</u>, with multiplicity 2I + 1.

MESONS	Y	I	Q	MASS (e.g.)
η	0	0 (G even)	0	η(549), ...
ϕ	0	0 (G odd)	0	ϕ(784) or , ϕ(1019), ...
ρ	0	1 (G even)	+1,0,-1	ρ(765), ...
π	0	1 (G odd)	+1,0,-1	π(140), ...
K	+1	1/2	+1,0	K(494), K(892), ...
$\bar{K}$	-1	1/2	0,-1	

BARYONS				
N	+1	1/2	+1,0	N(938), N(1670), ...
Δ	+1	3/2	+2,+1,0,-1	Δ(1238), Δ(1950), ...
Λ	0	0	0	Λ(115), Λ(1405),...
Σ	0	1	+1,0,-1	Σ(1192), Σ(1760), ...
Ξ	-1	1/2	0,-1	Ξ(1320), Ξ(1530), ...
Ω	-2	0	-1	Ω(1673), ...
Z_I	2	0,1	0 / +2,+1,0	

References

Particle Data Group. Review of Particle Properties. <u>Rev. Mod. Physics</u>, <u>43</u>, 51(1971).

2

INTRODUCTION TO UNITARY SYMMETRIES

We will attempt to introduce the classification of elementary particles which is based on the Unitary Symmetries. These are represented by groups of unitary unimodular transformations which leave the Lagrangian invariant. We recall that it is precisely this invariance under a given transformation, which leads to a particular conservation law. We will assume familiarity with the symmetries which lead to conservation of energy-momentum, of angular momentum, parity, to change configuration invariance, to invariance under time reversal.

Although through the study of Nuclear Physics, the concept of charge independence of nuclear forces should be also familiar, and its related isospin formalism, we will start with a review of the group SU(2), which deals with isospin, as a convenient introduction to the higher symmetry groups SU(3) and SU(6). Although a good deal of the mathematical ingredients necessary for this introduction are already familiar from quantum mechanics, in particular angular momentum, there is a language which made its way from group theory to elementary particle physics, which may be worth reviewing. Thus we will start with the elementary concepts of group theory.

Elements of Group Theory

We will summarize here some of the concepts of group theory which are relevant to the symmetries of elementary particles.

Definition of Group

A group is a set of elements $a, b, \ldots$ for which a composition rule, called product, exists, subject to the conditions:

1) The product of two elements $a, b,$ of the set is also an element of the set:

$$c = ab \qquad 2.1$$

(e.g., two rotations are equivalent to a third).

2) There exists a unit element e such that for any element a of the set:

$$ae = ea = a \qquad 2.2$$

(e.g., the unit rotation e leaves the system unchanged).

3) There exists, for each element a of the set, one and only one element a^{-1} such that

$$a^{-1}a = aa^{-1} = e \qquad 2.3$$

(e.g., Corresponding to each rotation, there is the inverse rotation, which restores the original condition.)

4) The product is associative

$$a(bc) = (ab)c \qquad 2.4$$

If, for all elements of a group

$$ab = ba \qquad 2.5$$

the group is said to be Abelian. (E.g., rotations about an axis form an Abelian group.)

-- There are discrete groups such as reflections and continuous groups such as rotations, etc.

-- If the elements of a continuous group contain a finite number of continuously varying parameters, the group is said to be finite. (E.g., Euler angles for rotations.)

-- If the parameters run over a finite range, like for rotations, the group is said to be compact.

-- A subset of a group is called a subgroup if the above properties hold for all its elements.

-- A subgroup is invariant if each of its elements commutes with each element of the complete group.

-- A group is simple if it contains no invariant subgroups, except the unit element.

Representation of the group:

Suppose that to each element of a group a, there corresponds a matrix $M(a)$ such that

$$M(a)M(b) = M(c), \quad \text{if} \quad ab = c$$

$$M(a^{-1}) = M^{-1}(a) \qquad 2.6$$

and

$$M(e) = I \qquad \text{(unit matrix)}.$$

These matrices and the vectors on which they act form a representation of the group. The dimension of the matrices in a representation is called the dimension of the representation. Under the operations of the group, a typical basis vector ξ_a transforms as

$$\xi_a \to \xi'_a = M^b_a \xi_b \qquad 2.7$$

with $a = 1,2,\ldots,n$.

Since the matrices M are required to have inverses (non-singular), in correspondence to every representation like ξ_a, there exists a contravariant representation $\bar{\xi}_a$ for which

$$\bar{\xi}^a \to \bar{\xi}'^a = \bar{\xi}^b (M^{-1})^a_b \qquad 2.8$$

Two representations $M_1(a)$ and $M_2(a)$ are equivalent if for every a there exists a constant matrix U such that

$$UM_1(a)U^{-1} = M_2(a) \qquad 2.9$$

Irreducible Representations

A representation is said to be completely reducible if every matrix M can be expressed or reduced to block diagonal form:

$$M = \begin{pmatrix} \boxed{M(1)} & & \\ & \boxed{M(2)} & \\ & & \boxed{M(k)} \end{pmatrix} \qquad 2.10$$

which can be written symbolically:

$$M = M(1) + M(2) + \ldots M(k) \qquad 2.11$$

If such reduction is not possible, the representation is called irreducible. Will specialize, give rules and examples in connection with the unitary groups later.

Direct Product Representation

Given two representations of a group $M^{\alpha}(1)$, $M^{\beta}(2)$ of dimensions α, β, one can obtain an $\alpha\beta$-dimensional representation of the group by taking the <u>direct</u> (or <u>Kronecker</u>) product of the corresponding matrices:

$$M^{\alpha\beta} = M^{\alpha}(1) \times M^{\beta}(2) \qquad 2.12$$

The direct product representation $M^{\alpha\beta}$ is completely reducible and can be split into a number of irreducible representations of dimension $\gamma_1, \gamma_2, \ldots, \gamma_n$.

$$M^{\alpha\beta} = M^{\gamma_1} + M^{\gamma_2} + \ldots M^{\gamma_n} \qquad 2.13$$

where

$$\gamma_1 + \gamma_2 + \ldots \gamma_n = \alpha\beta \qquad 2.14$$

The basis vector of the $M^{\alpha\beta}$ representation is obtained by the product of the basis vectors $\xi^{\alpha}(i)$, $\xi^{\beta}(k)$ of M^{α} and M^{β}:

$$\xi^{\alpha\beta}(ik) = \xi^{\alpha}(i) \cdot \xi^{\beta}(k) \qquad 2.15$$

The $\xi^{\alpha\beta}$ represent vectors in an $\alpha\beta$-dimensional linear space, or <u>tensors</u>.

Lie Groups

A group of transformations, which are characterized by a set of continuous parameters, is called a <u>Lie group</u>. We will consider only simple compact Lie groups. The group 0_3 of rotations in three dimensions is an example, i.e., the group of continuous transformations which leave invariant the real quadratic form

$$x_1^2 + x_2^2 + x_3^2 = \text{invariant} \qquad 2.16$$

as is well known, these transformations are expressed in terms of 3×3 orthogonal matrices. If the quadratic form 2.16 is generalized to n-dimension, we have the definition of the group 0_n.

The Unitary group U_n is the set of continuous transformations which leaves invariant the form

$$|\xi_1|^2 + |\xi_2|^2 + \ldots |\xi_n|^2 = \text{invariant} \qquad 2.17$$

The set is composed of the $n \times n$ unitary matrices which operate on the n-dimensional vector with complex components $\xi_1, \xi_2, \ldots, \xi_n$. We consider here the unitary transformations

$$U = e^{i\varepsilon^i F_i} \qquad (i = 1, \ldots, N) \qquad 2.18$$

which are represented by unitary operators, or unitary matrices:

$$U^\dagger U = UU^\dagger = 1; \quad U^{-1} = U^\dagger \qquad 2.19$$

The condition 2.19 also implies that the matrices F be Hermitian:

$$F_i = F_i^+ \qquad 2.20$$

Each transformation given by 2.18 is an element of a Lie group.

An infinitesimal transformation can be written as:

$$U = 1 + i\varepsilon^i F_i \qquad 2.21$$

or for a matrix element

$$U_{jk} = \delta_{jk} + i\varepsilon^i (F_i)_{jk} \qquad 2.22$$

The unitary infinitesimal transformations are unimodular if

$$\det \ U = 1 \qquad 2.23$$

From 2.22 this implies

$$\sum_j (F_i)_{jj} = 0 \quad \text{or} \quad \mathrm{Tr}\, F_i = 0 \qquad 2.24$$

namely the operators F_i are represented by traceless matrices. The operators F_i are called generators of the infinitesimal transformations or generators of the group. The commutation relations between these generators completely specify or characterize the "structure" of the group.

$$[F_i, F_j] = i\, f_{ij}^k\, F_k \qquad 2.25$$

The commutation relations 2.25 specify the algebra of the group. The coefficients f_{ij}^k are called the structure constants.

The number r of linearly independent generators F_i which characterize all the elements of the group, is called the order of the group. If there are ℓ independent operators F_i which commute, the number ℓ is called the rank of the group. If there are no commuting pairs of operators, the rank of the group is one.

E.g., $[J_i, J_j] = i\varepsilon_{ij}^k J_k$ for angular momentum operators rank one, take J_3 diagonal.

The Unitary Groups U(n) and SU(n)

The group $U(n)$ is represented by the transformations of an n-component complex spinor, by means of $(n \times n)$ unitary matrices. As in 2.7

$$\xi'_a = U_a^b \xi_b$$

where the unitary condition means

$$U^\dagger = U^{-1} \qquad U_b^{\dagger a} = (U_a^b)^* \qquad 2.27$$

Taking the complex conjugate of 2.26 we get

$$\xi_a'^* = \xi_b^* \, U_b^{\dagger a} \qquad 2.28$$

If we now define

$$\bar{\xi}^a \equiv (\xi_a)^* \qquad 2.29$$

from 2.27, and 2.28 we have

$$\bar{\xi}'^a = \bar{\xi}^b \, U^{-1a}_{\ \ b} \qquad 2.30$$

in accord with 2.8. Check if 2.17 is satisfied:

$$|\xi_1|^2 + |\xi_2^2 + \ldots = \bar{\xi}^a \xi_a \to \bar{\xi}'^a \xi'_a = \bar{\xi}^b \underbrace{U^{-1a}_{\ \ b} \, U_a^b}_{=1} \xi_b =$$

$$\bar{\xi}\xi = \underline{\text{invariant}}.$$

If the transformations are <u>unimodular</u> (Det U = 1), the group is called SU(n). As already seen above, essentially

$$U = e^{iH}; \; U_a^b = e^{iH_a^b}; \; \text{Det } U = 1; \; \text{Tr } H = H_a^a = 0. \qquad 2.31$$

<u>References</u>: See References following Lecture 4.

3

THE GROUP SU(2)

The group describes the transformations of spinors with two components. It is related to isotopic spin and to ordinary angular momentum. Will describe first the general properties of the group, then we will make the association of the basis vectors and irreducible representations with specific particle states.

Consider a two-component complex vector (spinor) x and its complex conjugate x^*:

$$x = \begin{pmatrix} x_1 \\ x_2 \end{pmatrix}, \quad x^* = \begin{pmatrix} x_1^* \\ x_2^* \end{pmatrix} \equiv \begin{pmatrix} \bar{x}^1 \\ \bar{x}^2 \end{pmatrix} = \bar{x} \qquad 3.1$$

x can be regarded either as the particle state vector or the field operator which annihilates the particle. Corresponding to the latter, the complex conjugate operator annihilates antiparticles. Thus $x^* \equiv \bar{x}$ can be regarded as describing the antiparticle state vector. The most general unitary unimodular transformation $x' = Ux$ 3.2

is represented by 2×2 matrices

$$U = \begin{pmatrix} \alpha & \beta \\ -D\beta^* & D\alpha^* \end{pmatrix}, \quad \text{with } \alpha\alpha^* + \beta\beta^* = 1, \quad D = \det U, \ |D| = 1 \qquad 3.3$$

As in 2.18, a unitary transformation can be written in terms of a set of linearly independent Hermitian matrices. There are four 2×2 such matrices, for instance the Paul's matrices:

$$\tau_1 = \begin{pmatrix} 0 & 1 \\ 1 & 0 \end{pmatrix}, \quad \tau_2 = \begin{pmatrix} 0 & -i \\ i & 0 \end{pmatrix}, \quad \tau_3 = \begin{pmatrix} 1 & 0 \\ 0 & -1 \end{pmatrix} \qquad 3.4$$

plus the unit matrix $\mathbb{1} = \begin{pmatrix} 1 & 0 \\ 0 & 1 \end{pmatrix}$. Any 2×2 Hermitian matrix can be expressed as a linear combination of these four matrices, or that a generical unitary transformation is given by

$$U = e^{i\varepsilon_0 \mathbb{1}} \cdot e^{i\vec{\varepsilon}\cdot\vec{\tau}} \qquad 3.5$$

where the τ_i have been written formally as components of a 3-vector

$$\vec{\tau} \equiv (\tau_1, \tau_2, \tau_3) \qquad 3.6$$

and $\vec{\varepsilon}$ is a 3-vector with real components $\varepsilon_1, \varepsilon_2, \varepsilon_3$.

The factor $e^{i\varepsilon_0 \mathbb{1}}$ gives a "phase transformation," which leaves the basis vector unchanged (e.g., baryon conservation). Although this transformation is contained in U(2), SU(2), because of the traceless condition, reduces to

$$U = e^{i\vec{\varepsilon}\cdot\vec{\tau}} \quad \text{or} \quad U = e^{i\vec{\varepsilon}\cdot\frac{\vec{\tau}}{2}} \quad \text{or} \quad U = \mathbb{1} + i\vec{\varepsilon}\cdot\vec{\tau}/2. \qquad 3.7$$

The operators $I_i = \dfrac{\tau_i}{2}$ 3.8

are the infinitesimal generators of the group SU(2). I is called <u>isotopic spin</u>. The transformations 3.7 define the group SU(2); they correspond to its irreducible representation of dimension 2. The <u>order</u> of the group is 3, since there are three independent generators τ_1, τ_2, τ_3.

The algebra of the group is defined by the commutation relations

$$[I_i, I_j] = i\,{}_{ij}^{k} I_k \qquad 3.9$$

Since there is no commuting pair of generators, the rank of the group is <u>one</u>. The only diagonal generator is I_3 with eigenvalues +1/2 and -1/2. Clearly the group structure is identical to that describing ordinary spin angular momentum. Accordingly we can define raising and lowering or step-operators

$$\tau_{\pm} = \frac{1}{2}(\tau_1 \pm i\tau_2) \qquad 3.10$$

$$\tau_+ = \begin{pmatrix} 0 & 1 \\ 0 & 0 \end{pmatrix} ; \quad \tau_- = \begin{pmatrix} 0 & 0 \\ 1 & 0 \end{pmatrix} \qquad 3.11$$

which operate on the base vectors, e.g.

$$x_+ = \begin{pmatrix} 1 \\ 0 \end{pmatrix} \text{, up} \quad \text{and} \quad x_- = \begin{pmatrix} 0 \\ 1 \end{pmatrix} \text{, down} \qquad 3.12$$

yielding

$$\tau_{\pm} x_{\pm} = 0 ; \quad \tau_{\pm} x_{\mp} = x_{\pm} \qquad 3.13$$

The set of matrices $\tau_+, \tau_-, \tau_3, \mathbb{1}$, is another alternative representation of the group, with generators I_+, I_-, I_3. The commutation relations become

$$[I_3, I_{\pm}] = \pm I_{\pm} \qquad 3.14$$

The eigenvalues of I_3, $\pm\frac{1}{2}$ are also called <u>weights</u>. All the mathematical apparatus of spin angular momentum can be transferred to the isospin case. Relevant results are, in summary

$$[I_1, I_2] = iI_3; \quad I_3|I,M\rangle = M|I,M\rangle,$$

$$I^2|I,M\rangle = I(I+1)|I,M\rangle; \quad [I^2, I\pm] = 0 \qquad 3.15$$

$$I\pm|I,M\rangle = \sqrt{I(I+1)-M(M\pm1)}\ J,M\pm1\rangle$$

Product Representations and Irreducible Representations of SU(2)

Each irreducible representation is characterized by the "dominant weight" $I = \lambda\frac{1}{2}$ ($\lambda=0,1,2,\ldots$) representing the maximum eigenvalue I_3. It is represented by $D(\frac{1}{2})$, and its dimension is $(2I+1)$. The basis vectors are $x(I,I_3)$. The two-dimensional representation is then denoted by $D(1/2)$. In order to streamline the notation for the basis vector, will adopt

$$x_1 \equiv x_+ \equiv x(\tfrac{1}{2},\tfrac{1}{2}) \quad ; \quad x_2 \equiv x_- \equiv x(\tfrac{1}{2},-\tfrac{1}{2}) \qquad 3.16$$

We are certainly familiar already with the result

$$D(1/2) \times D(1/2) = D(0) + D(1) \qquad 3.17$$

The dimensions of $D(0)$ and $D(1)$ are 1 and 3 respectively, and we can write symbolically

$$2 \times 2 = 1 + 3. \qquad 3.18$$

The basis for $D(0)$ and $D(1)$ are the usual eigenstates of isospin (or angular momentum) obtained from the antisymmetric and symmetric combinations of the $D(1/2)$ basis states.

$$D(1/2) \times D(1/2) = x_\alpha \cdot x_\beta = M_{\alpha\beta} = M_{[\alpha\beta]} + M_{\{\alpha\beta\}} \qquad 3.19$$

$$M_{[\alpha\beta]} = x(0,0) = \frac{1}{\sqrt{2}}(x_1x_2 - x_2x_1) = S \qquad 3.20$$

(scalar, invariant)

$$M_{\{\alpha\beta\}} = \left.\begin{array}{lll} x(1,1) & = x_1x_1 & = V^+ \\ x(1,0) & = \frac{1}{\sqrt{2}}(x_1x_2 - x_2x_1) & = V^0 \\ x(1,-1) & = x_2x_2 & = V^- \end{array}\right\} \text{vector} \qquad 3.21$$

The transformation properties can be verified by actually transforming the basis vectors by means of e.g., an infinitesimal transformation. It is found that $x(0,0)$ transforms into itself, and that $x(1,1)$, $x(1,0)$, $x(1,-1)$ transform into linear combinations of themselves. From the exponential form of the infinitesimal transformation 3.7 it follows that the <u>product</u> of two transformations corresponds to the <u>sum</u> of the generators. Then if $I(1)$, $I(2)$ are the isospins for two representations $D(1/2)$, the <u>total isospin</u>, corresponding to the product representation is

$$\vec{I} = \vec{I}(1) + \vec{I}(2) \qquad 3.22$$

Accordingly, it could be checked that the bases of the product representation $x(0,0)$, $x(1,\pm 1,0)$ corresponds to the eigenvalues of I^2, I_3. All of this is of course well known from the elementary treatment of angular momentum, it is simply reformulated here with the vocabulary of groups. The generalization of 3.17 or 3.18, to the direct product of irreducible representations of arbitrary dimensions is, in SU(2):

$$D[I(1)] \times D[I(2)] = D[I(1)+I(2)] + D[I(1)+I(2)-1] + \ldots D[|I(1)-I(2)|] \qquad 3.23$$

We can use a graphical method to combine two spinor states: symbolize the spinor state by a double arrow:

To construct the states of two spinors, place a second double arrow with its center on either of the two ends of the first.

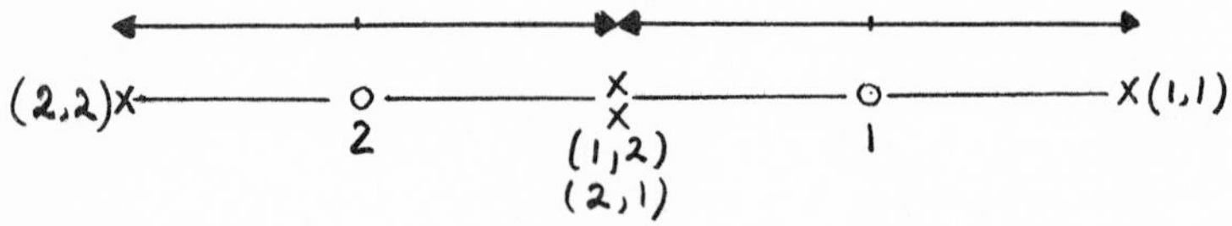

These are in fact, <u>weight diagrams</u>. The states (1,2) and (2,1) fall on the same spot. With these construct first the invariant (12 - 21) and then take the orthogonal combination to obtain the member of the triplet. This construction, which may seem trivial in SU(2), will become very useful in SU(3).

References: See references following Lecture 4.

4

APPLICATION OF SU(2) TO ELEMENTARY PARTICLES

We can classify elementary particles into <u>isospin multiplets</u>, by assigning each multiplet to an irreducible representation of SU(2). For example, we can identify the proton p and the neutron n with the basis vectors of the I. R. D(1/2)

$$\underset{\sim}{x} = \begin{pmatrix} p \\ n \end{pmatrix} = \begin{pmatrix} x_1 \\ x_2 \end{pmatrix} \qquad 4.1$$

in accord with the Gell-Mann Nishijima relation

$$Q = I_3 + \frac{1}{2} Y$$

As mentioned already (3.1) we identify the antiparticle states with x*

$$\underset{\sim}{x}^* = \begin{pmatrix} \bar{p} \\ \bar{n} \end{pmatrix} = \begin{pmatrix} \bar{x}^1 \\ \bar{x}^2 \end{pmatrix} \qquad 4.2$$

We have already seen that such states transform in a contravariant fashion (2.30). We can translate this property in a very simple rule as follows:

$$\begin{pmatrix} p \\ n \end{pmatrix} \quad \text{transforms like} \begin{pmatrix} \bar{n} \\ -\bar{p} \end{pmatrix} \qquad 4.3$$

under a unitary unimodular transformation. In fact, for example

$$\underset{\sim}{x}' = U\underset{\sim}{x} \qquad x' = \begin{pmatrix} \alpha & \beta \\ -\beta^* & \alpha^* \end{pmatrix} \begin{pmatrix} p \\ n \end{pmatrix} = \begin{pmatrix} \alpha p+\beta n \\ -\beta^* p+\alpha^* n \end{pmatrix} \quad 4.4$$

$$\underset{\sim}{x}'^* = U^*\underset{\sim}{x}^* \qquad x'^* = \begin{pmatrix} \alpha^* & \beta^* \\ -\beta & \alpha \end{pmatrix} \begin{pmatrix} \bar{p} \\ \bar{n} \end{pmatrix} = \begin{pmatrix} \alpha^*\bar{p}+\beta^*\bar{n} \\ -\beta\bar{p} +\alpha\bar{n} \end{pmatrix} \quad 4.5$$

As can be seen, $p \to \alpha p$, $\bar{n} \to \alpha\bar{n}$; $n \to \beta n$, $\bar{p} \to -\beta\bar{p}$ which illustrate 4.3.

In terms, then, of the particle states p,n, the isospin eigenstates 3.20, 3.21 become

$$\begin{aligned} x(0,0) &= \frac{1}{\sqrt{2}} (pn-np) \\ x(1,1) &= pp \\ x(1,0) &= \frac{1}{\sqrt{2}} (pn-np) \\ x(1,-1) &= nn \end{aligned} \qquad 4.6$$

and the charges Q obtain from $Q = I_3 + \frac{1}{2}Y$ with $Y = 2$ (two baryons). For a nucleon-antinucleon system on the other hand, there is a significant difference. Using the rule 4.3 to obtain eigenstates having the correct transformation properties, after an overall phase change we find,

$$\begin{aligned} x(0,0) &= \frac{1}{\sqrt{2}} (\bar{p}p+\bar{n}n) \\ x(1,1) &= \bar{n}p \\ x(1,0) &= \frac{1}{\sqrt{2}} (\bar{p}p-\bar{n}n) \\ x(1,-1) &= \bar{p}n \end{aligned} \qquad 4.7$$

Note that these eigenstates have $B = 0$, $Y \neq 0$, quantum numbers of the non-strange mesons. We expand on the $N\bar{N}$ system a little further, in particular to connect with the notions previously introduced about direct product representation. What we have done to obtain 4.7 is indeed the following: We have taken the <u>direct</u> or <u>outer</u> or <u>tensor</u> product

$$\bar{N} \otimes N = \bar{x}^i x_j \equiv M_\alpha^\beta = \begin{pmatrix} M_1^1 & M_1^2 \\ M_2^1 & M_2^2 \end{pmatrix} = \begin{pmatrix} \bar{p}p & \bar{n}p \\ \bar{p}n & \bar{n}n \end{pmatrix} \qquad 4.8$$

We can rewrite 4.8 as follows:

$$M_\alpha^\beta = \begin{pmatrix} \frac{\bar{p}p+\bar{n}n}{2} + \frac{\bar{p}p-\bar{n}n}{2} & \bar{n}p \\ \bar{p}n & \frac{\bar{p}p+\bar{n}n}{2} - \frac{\bar{p}p-\bar{n}n}{2} \end{pmatrix}$$

$$= \frac{1}{\sqrt{2}} \begin{pmatrix} \frac{\bar{p}p+\bar{n}n}{\sqrt{2}} & 0 \\ 0 & \frac{\bar{p}p+\bar{n}n}{\sqrt{2}} \end{pmatrix} + \frac{1}{\sqrt{2}} \begin{pmatrix} \frac{\bar{p}p-\bar{n}n}{\sqrt{2}} & \bar{n}p \\ \bar{p}n & -\frac{\bar{p}p-\bar{n}n}{\sqrt{2}} \end{pmatrix} \qquad 4.9$$

$$= \underbrace{\frac{1}{2}\delta_\alpha^\beta M_\gamma^\gamma}_{\substack{\text{scalar} \\ \text{one component}}} + \underbrace{M_\alpha^\beta - \frac{1}{2}\delta_\alpha^\beta M_\gamma^\gamma}_{\substack{\text{traceless tensor} \\ \text{three components} = 4 - 1}} \qquad 4.10$$

This tensor decomposition will be very useful for the construction of irreducible representations in SU(3), even if it may seem superfluous at this point. First of all we see that the scalar term can be considered <u>either</u> the trace of $\bar{N}^\beta N_\alpha$ or as the antisymmetric combination of

$$N^*_\beta N_\alpha = M_{[\beta\alpha]} \quad \text{where} \quad N^*_\beta = \begin{pmatrix} \bar{n} \\ -\bar{p} \end{pmatrix} . \qquad 4.11$$

The components of M^β_α can be made to correspond to the pseudoscalar meson field operators. In particular we can identify the scalar with the η-meson, isospin singlet

$$\eta = \frac{\bar{p}p+\bar{n}n}{\sqrt{2}} \qquad 4.12$$

and the vector components with the pion triplet:

$$\pi^0 = \frac{\bar{p}p-\bar{n}n}{\sqrt{2}} , \quad \pi^+ = \bar{n}p, \quad \pi^- = \bar{p}n \qquad 4.13$$

Note that the charge labels correspond to the components I_3 of the $\bar{N}N$ combinations. Alternative labelling of the π components is

$$\pi^0 = \pi^3; \quad \pi^\pm = \frac{\pi^1 \mp i\pi^2}{\sqrt{2}} \qquad 4.14$$

Thus

$$\bar{N} \otimes N = \eta\frac{1}{\sqrt{2}} \mathbb{1} + \begin{pmatrix} \frac{\pi^0}{\sqrt{2}} & \pi^+ \\ \pi^- & -\frac{\pi^o}{\sqrt{2}} \end{pmatrix} \qquad 4.15$$

or using 4.14

$$= \eta\frac{1}{\sqrt{2}} \mathbb{1} + \frac{1}{\sqrt{2}} \vec{\tau}\cdot\vec{\pi} \qquad 4.16$$

Since $\tau_\pm = \frac{\tau_1+i\tau_2}{2}$, we also have

$$\frac{1}{\sqrt{2}} \vec{\tau}\cdot\vec{\pi} = \frac{1}{\sqrt{2}} \pi^3\tau_3 + \pi^+\tau_+ + \pi^-\tau_- \qquad 4.17$$

At this point the classification of the "stable" (under strong interaction) particles from the point of view of SU(2) is essentially complete. We have baryons of spin $\frac{1}{2}$ in form of doublets and a "quartet"

$$N = \begin{pmatrix} p \\ n \end{pmatrix}, \quad \Xi = \begin{pmatrix} \Xi^0 \\ \Xi^- \end{pmatrix} \qquad \Sigma = (\Sigma^+, \Sigma^0, \Sigma^-), \ \Lambda \qquad 4.18$$

The mesons of spin zero are similarly:

$$K = \begin{pmatrix} K^+ \\ K^0 \end{pmatrix}, \quad \bar{K} = \begin{pmatrix} \bar{K}^0 \\ K^- \end{pmatrix}; \qquad \pi = (\pi^+, \pi^0, \pi^-), \ \eta \qquad 4.19$$

The existence of Ξ^0, Σ^0, and of distinct K^0 and $\bar{K}^0$ was predicted by Gell-Mann and by Nishijima and Nakau in 1953, on the basis of this isospin classification.

Caution: The construction of the representation for the pions which has been given as a detailed example could be given a physical basis in terms of nucleon-antinucleon states. In fact this was the idea of the Fermi-Yang model. Once a representation is formally constructed however, it is of course applicable to any multiplet having the same transformation properties. Thus one can represent the baryon quartet formally as

$$M_\alpha^\beta = \Lambda \frac{1}{\sqrt{2}} \mathbb{1} + \begin{pmatrix} \frac{\Sigma^0}{\sqrt{2}} & \Sigma^+ \\ \Sigma^- & -\frac{\Sigma^0}{\sqrt{2}} \end{pmatrix} \qquad 4.20$$

without implying particular bound states. By combining a pion with a nucleon, six different states are obtained.

$$D(1) \times D(1/2) = D(1/2) + D(3/2) \qquad 4.21$$

$$3 \otimes 2 \qquad\qquad 2 \oplus 4$$

This result can be obtained in various ways, using for example the standard rules for the addition of angular momenta, Clebsch-Gordan coefficients, etc. The construction of the eigenstates for this representation has been illustrated in gory detail, using the appropriate shift operators, in Levi Setti (1962). One could now arrive at the same result using the tensor method. In fact the problem reduces to find the irreducible representations of the tensor

$$\pi_{\alpha}^{\beta} N_{\gamma} = M_{\alpha\gamma}^{\beta} \qquad 4.22$$

and their components. This however will be left as an exercise.

The experimental justification for isospin does not limit itself of course to the classification of particles in charge multiplets. Several implications of charge independence in particle reactions have been discussed in Levi Setti (1962). To conclude this brief review of SU(2), we can just summarize: Charge independence of strong interactions is equivalent to conservation of the total isospin. The third component is also conserved, giving charge conservation. Thus for the two-nucleon system, instead of four possible interactions (corresponding to the four components of $M_{\alpha\beta}$), there are only two, the singlet and the triplet interactions. Similarly for the pion-nucleon system, the strong interaction distinguishes only between $I = 1/2$ and $I = 3/2$ states. In fact we see that the number of different amplitudes describing the interaction equals the number of irreducible representations into which the product representation of the interacting particles can be decomposed.

A word about the Yukawa interaction: Invariance of the S-matrix under isospin transformations requires that the scattering amplitudes must be scalar under such transformations. The Yukawa interaction is described by the product of the $\bar{N}$, N and π fields $\bar{N}^a N_b \phi_c$. Since ϕ_c transforms like a vector, to obtain a scalar, also $\bar{N}N$ must transform like a vector, namely all suffices should be saturated. But this can be achieved very simply by taking the vector part of our field from 4.16, so that the interaction energy will take the form

$$g\bar{N}^{\alpha}\pi^{\beta}_{\alpha}N_{\beta} = \frac{g}{\sqrt{2}}\,\bar{N}^{\alpha}(\tau_i)^{\beta}_{\alpha}N_{\beta}\pi^{i} = \frac{g}{\sqrt{2}}\,\bar{N}\vec{\tau}N\cdot\vec{\pi} \qquad 4.23$$

and from 4.14 and 4.17:

$$\frac{g}{\sqrt{2}}\,\bar{N}\vec{\tau}_N\cdot\bar{\pi} = \frac{g}{\sqrt{2}}\,(\bar{p}p-\bar{n}n)\pi^{0} + g\bar{p}n\pi^{+} + gp\bar{n}\pi^{-}. \qquad 4.24$$

References:

Much of the presentation in Lectures 2, 3 and 4 is based on the review by P.T. Matthews, Unitary Symmetry, in High Energy Physics, Vol. 1. Edited by E.H.S. Burhop. Academic Press, New York and London, 1967.

Also relevant have been lecture notes by:

G. Costa, Introduction to Unitary Symmetries. Proceedings 1964 Herceg-Novi Easter School. CERN Report 64-13.

R. Levi Setti, Elementary Particles. The University of Chicago Press. Chicago and London, 1963.

5

THE GROUP SU(3)

Quarks

In order to incorporate conservation of strangeness or hypercharge together with isospin conservation within a group structure for strong interactions, we need a group of rank two. (As previously remarked, such a group will give two additive quantum numbers.) The required group is provided by extending SU(2) to SU(3), namely the group of unitary unimodular transformations of dimension 3. The ingredients for doing so are for the most part already available from what has been done for SU(2). The main difference consists in the fact that our basic spinors now have 3-components rather than 2. We define

$$q_i = \begin{pmatrix} p \\ n \\ \lambda \end{pmatrix} \quad ; \quad \bar{q}^i = (\bar{p}, \bar{n}, \bar{\lambda}) \qquad 5.1$$

and associate to this spinor the charge matrix (the justification for these choices will become apparent later).

$$Q = \begin{pmatrix} 2/3 & & \\ & -1/3 & \\ & & -1/3 \end{pmatrix} \quad \text{and} \quad I_3 = \begin{pmatrix} 1/2 & & \\ & -1/2 & \\ & & 0 \end{pmatrix} \qquad 5.2$$

We have added to the two basic states of SU(2), p, n, of isospin 1/2, a third state λ of isospin 0, carrier of strangeness $S = -1$. To satisfy the Gell-Mann Nishijima relation, the assignments 5.2 also imply that

$$B = \begin{pmatrix} 1/3 & & \\ & 1/3 & \\ & & 1/3 \end{pmatrix} \qquad 5.3$$

Gell-Mann called these entities <u>quarks</u>. (M. Gell-Mann, Physics Letters 8,214(1964)).

These are the quantum numbers of q_i.

	I_3	Y	Q	B
p	+1/2	1/3	2/3	1/3
n	-1/2	1/3	-1/3	1/3
λ	0	-2/3	-1/3	1/3

5.4

For $\bar{q}^i$, the antiquarks, all entries in the table have the opposite sign. The quarks can be represented as <u>weight diagrams</u> in the $I_3 - Y$ plane

5.5

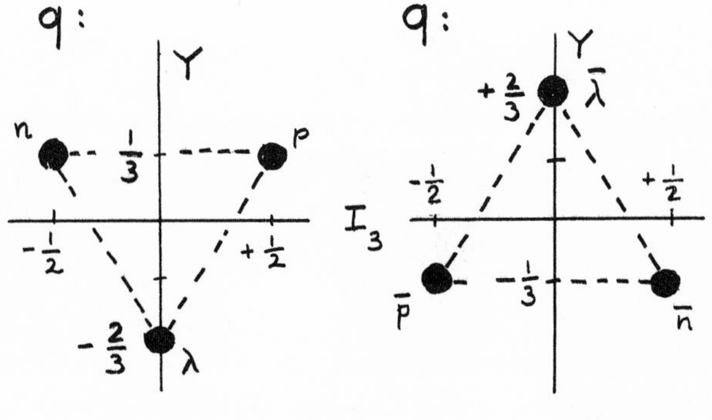

Note the symmetry of these diagrams. There are 3-axes of symmetry, and with respect to each axis, the triplet may be taken as being made of a doublet plus a singlet. One can in fact define three SU(2) doublets, corresponding to I-spin, U-spin and V-spin: (p,n), (λ,n) and (p,λ) respectively.

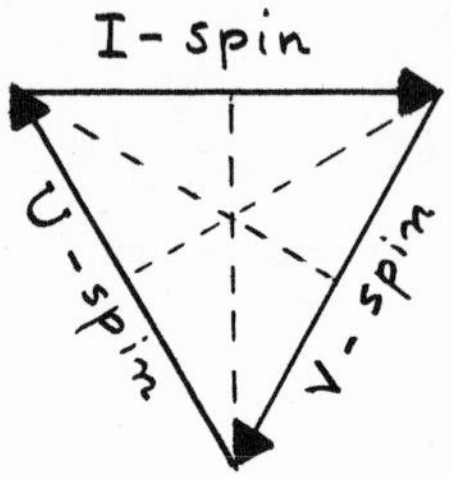

5.6

Product Representation of Quark and Antiquark

Perhaps the most instructive introduction to the formalism of SU(3) is to consider in some detail the representations which can be constructed with q and $\bar{q}$. A very intuitive approach is to use the same graphical construction seen in SU(2). To do this we just superimpose the center of the quark triangle (or antiquark) over each state of the other quark (or antiquark) in the product representation. For example

$q\bar{q}$:

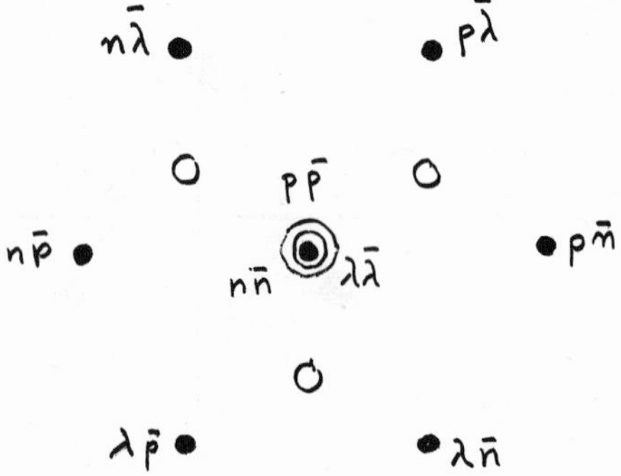

$3 \otimes \bar{3} = 1 \oplus 8$

Out of the three states at the center, we will construct proper linear combinations, orthogonal among each other, and of defined symmetry, as for SU(2). In fact we are going to do this right now, taking advantage of the tensor decomposition just seen for SU(2). Our $q\bar{q}$ tensor is now

$$q \otimes \bar{q} = q_i \bar{q}^j = M^{\beta}_{\alpha} = \begin{pmatrix} p\bar{p} & p\bar{n} & p\bar{\lambda} \\ n\bar{p} & n\bar{n} & n\bar{\lambda} \\ \lambda\bar{p} & \lambda\bar{n} & \lambda\bar{\lambda} \end{pmatrix} \qquad 5.7$$

The extension of 4.10 to $\alpha,\beta = 1,2,3$ will be

$$M^{\beta}_{\alpha} = \underbrace{\frac{1}{3}\delta^{\beta}_{\alpha} M^{\gamma}_{\gamma}}_{\substack{\text{scalar} \\ \underline{\text{singlet}}}} + \underbrace{M^{\beta}_{\alpha} - \frac{1}{3}\delta^{\beta}_{\alpha} M^{\gamma}_{\gamma}}_{\substack{\text{traceless tensor} \\ 8 \text{ components } = q - 1 = \underline{\text{octet}}}} \qquad 5.8$$

For the singlet we have then

$$\{1\} = \frac{1}{3}(p\bar{p}+n\bar{n}+\lambda\bar{\lambda}) = \frac{1}{\sqrt{3}} X \qquad 5.9$$

if we call

$$X = \frac{p\bar{p}+n\bar{n}+\lambda\bar{\lambda}}{\sqrt{3}} \qquad 5.10$$

If we want to assign this state to one of the 0^- mesons, a good candidate is in fact the X-meson X(958) or η'. For the octet

$$\{8\} = \begin{pmatrix} \frac{1}{3}(2p\bar{p}-n\bar{n}-\lambda\bar{\lambda}) & p\bar{n} & p\bar{\lambda} \\ n\bar{p} & \frac{1}{3}(-p\bar{p}+2n\bar{n}-\lambda\bar{\lambda}) & n\bar{\lambda} \\ \lambda\bar{p} & \lambda\bar{n} & \frac{1}{3}(-p\bar{p}-n\bar{n}+2\lambda\bar{\lambda}) \end{pmatrix} \qquad 5.11$$

We can manipulate the diagonal elements as follows:

$$\frac{2\bar{p}p-\bar{n}n-\bar{\lambda}\lambda}{3} = \frac{\frac{1}{2}(p\bar{p}+n\bar{n})+\frac{3}{2}(p\bar{p}-n\bar{n})-\lambda\bar{\lambda}}{3} = \frac{p\bar{p}+n\bar{n}-2\lambda\bar{\lambda}}{6} + \frac{1}{2}(p\bar{p}-n\bar{n})$$

$$= \frac{\eta}{\sqrt{6}} + \frac{\pi^0}{\sqrt{2}} \qquad 5.12$$

where we define

$$\eta = \frac{p\bar{p}+n\bar{n}-2\lambda\bar{\lambda}}{\sqrt{6}} \qquad 5.13$$

(This is the same η-meson, encountered in SU(2), now made out of $q\bar{q}$) and $\pi^0 = \frac{p\bar{p}-n\bar{n}}{\sqrt{2}}$ is still the same old π^0 of SU(2).

What particles can we associate to the other matrix elements of M_α^β? To $p\bar{n}$ and $n\bar{p}$, the π^+ and π^- respectively, as in SU(2). With the quantum number assignments to the quarks of 5.2 we see that

$$p\bar{\lambda} \equiv K^+ \qquad n\bar{\lambda} \equiv K^0$$
$$\lambda\bar{p} \equiv K^- \qquad \lambda\bar{n} \equiv \bar{K}^0 \qquad 5.14$$

Thus for the 0^- mesons

$$M_\alpha^\beta = \frac{1}{\sqrt{3}} \times \mathbb{1} + \begin{pmatrix} \frac{\eta}{\sqrt{6}} + \frac{\pi^0}{\sqrt{2}} & \pi^+ & K^+ \\ \pi^- & \frac{\eta}{\sqrt{6}} - \frac{\pi^0}{\sqrt{2}} & K^0 \\ K^- & \bar{K}^0 & -\frac{2\eta}{\sqrt{6}} \end{pmatrix} \qquad 5.15$$

We have thus constructed the proper combinations of the neutral components of $q\bar{q}$ to assign to the triple occupancy at the center of the diagram constructed on page 32.

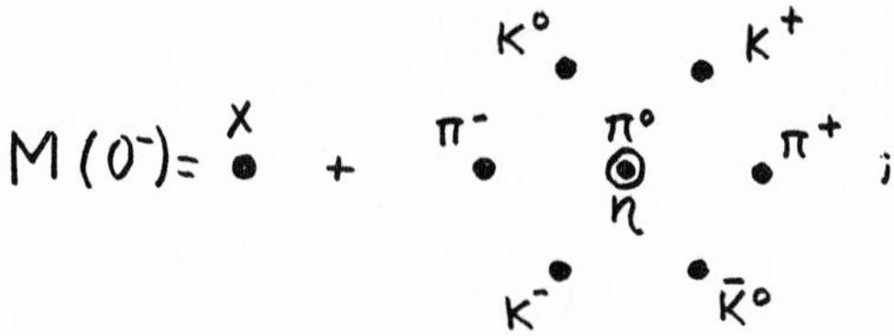

$$M(0^-) = \overset{x}{\bullet} \; + \quad \begin{matrix} & \overset{K^o}{\bullet} & & \overset{K^+}{\bullet} & \\ \overset{\pi^-}{\bullet} & & \overset{\pi^o}{\underset{\eta}{\circledcirc}} & & \overset{\pi^+}{\bullet} \\ & \overset{K^-}{\bullet} & & \overset{\bar{K}^o}{\bullet} & \end{matrix} \; ; \qquad 5.16$$

This is the so-called 0^--meson <u>nonet</u>.

Another well-known meson nonet is represented by the 1^- or vector mesons:

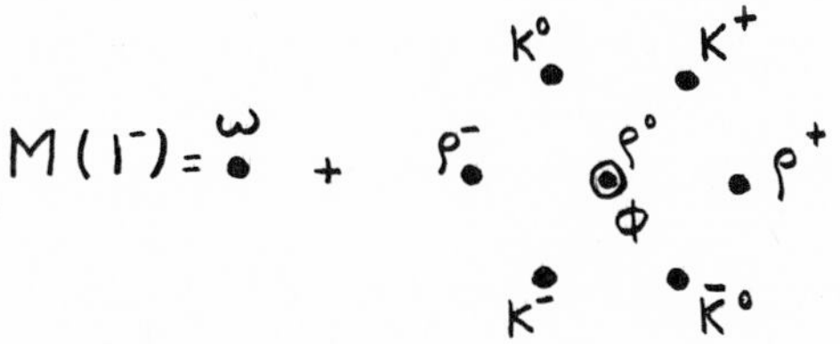

$$M(1^-) = \overset{\omega}{\bullet} \; + \quad \begin{matrix} & \overset{K^o}{\bullet} & & \overset{K^+}{\bullet} & \\ \overset{\rho^-}{\bullet} & & \overset{\rho^o}{\underset{\phi}{\circledcirc}} & & \overset{\rho^+}{\bullet} \\ & \overset{K^-}{\bullet} & & \overset{\bar{K}^o}{\bullet} & \end{matrix} \qquad 5.17$$

where K stands for $K(890)$ or K^*.

Generators of SU(3)

Having already obtained the structure of the octet representation, we can relate such structure to the generators of SU(3) in a very straightforward manner.

Remember how, in SU(2), we have related the π-matrix to the generators τ_i:

$$\begin{pmatrix} \frac{\pi^0}{\sqrt{2}} & \pi^+ \\ \pi^- & -\frac{\pi^o}{\sqrt{2}} \end{pmatrix} = \frac{1}{\sqrt{2}} \begin{pmatrix} \pi^3 & \pi^1 - i\pi^2 \\ \pi^1 + i\pi^2 & -\pi^3 \end{pmatrix} \qquad 5.18$$

$$= \frac{1}{\sqrt{2}} (\tau_i)^b_a \pi^i = \frac{1}{\sqrt{2}} \vec{\tau}\cdot\vec{\pi}.$$

Now, we could have constructed the SU(3) octet 5.15 using a relation completely similar to 5.18,

$$\{8\} = \frac{1}{\sqrt{2}} (\lambda_i)^b_a M^i = \frac{1}{\sqrt{2}} \vec{\lambda}\cdot\vec{M} \qquad 5.19$$

where $\vec{M}$ is an object with 8 components, an octet, and λ_i is a set of 3×3 matrices entirely similar to the τ_i in SU(2).

The infinitesimal transformations will be here

$$U = \mathbb{1} + i\,\vec{\varepsilon}\cdot\frac{\vec{\lambda}}{2} \quad \text{or} \quad U = \mathbb{1} + i\,\vec{\varepsilon}\cdot\vec{F} \qquad 5.20$$

where $F_i = \frac{\lambda_i}{2}$ 5.21 are the infinitesimal generators of the group. The λ-matrices are

$$\lambda_1 = \left(\begin{array}{cc|c} 0 & 1 & 0 \\ 1 & 0 & 0 \\ \hline 0 & 0 & 0 \end{array}\right) \qquad \lambda_2 = \left(\begin{array}{cc|c} 0 & -i & 0 \\ i & 0 & 0 \\ \hline 0 & 0 & 0 \end{array}\right) \qquad \lambda_3 \left(\begin{array}{cc|c} 1 & 0 & 0 \\ 0 & -1 & 0 \\ \hline 0 & 0 & 0 \end{array}\right)$$

$$\lambda_4 = \begin{pmatrix} 0 & 0 & 1 \\ 0 & 0 & 0 \\ 1 & 0 & 0 \end{pmatrix} \qquad \lambda_5 = \begin{pmatrix} 0 & 0 & -i \\ 0 & 0 & 0 \\ i & 0 & 0 \end{pmatrix} \qquad \lambda_6 \begin{pmatrix} 0 & 0 & 0 \\ 0 & 0 & 1 \\ 0 & 1 & 0 \end{pmatrix}$$

$$\lambda_7 = \begin{pmatrix} 0 & 0 & 0 \\ 0 & 0 & -i \\ 0 & i & 0 \end{pmatrix} \qquad \lambda_8 = \sqrt{1/3} \begin{pmatrix} 1 & 0 & 0 \\ 0 & 1 & 0 \\ 0 & 0 & -2 \end{pmatrix} \qquad 5.22$$

and the $F_i = \frac{\lambda i}{2}$ obey the commutation relations

$$|F_i, F_j| = i\, f_{ij}^{k}\, F_k \qquad 5.23$$

and the anticommutation relations

$$\{F_i, F_j\} = \frac{1}{3}\delta_{ij} + d_{ij}^{k}\, F_k \qquad 5.24$$

The f_{ij}^k and d_{ij}^k are the structure constants of SU(3), usually tabulated. Before going further, we complete the connection with the octet matrix. Using 5.19 and 5.22, the result found in 5.15 can be retranscribed as

$$\begin{pmatrix} \frac{\pi^o}{\sqrt{2}}+\frac{\eta}{\sqrt{6}} & \pi^+ & K^+ \\ \pi^- & -\frac{\pi^o}{\sqrt{2}}+\frac{\eta}{\sqrt{6}} & K^o \\ K^- & \bar{K}^o & -\frac{2\eta}{\sqrt{6}} \end{pmatrix} \equiv \frac{1}{\sqrt{2}} \begin{pmatrix} M^3+\frac{M^8}{\sqrt{3}} & M^1 iM^2 & M^4-iM^5 \\ M^1+iM^2 & -M^3+\frac{M^8}{\sqrt{3}} & M^6-iM^7 \\ M^4+iM^5 & M^6+iM^7 & -2\frac{M^8}{\sqrt{3}} \end{pmatrix} \qquad 5.25$$

We could have of course started to construct the octet from this end; the correspondence of ten matrix elements to actual particle states would have been, however, much less intuitive.

Note that $\lambda_1, \lambda_2, \lambda_3$ are just the Pauli matrices with an additional column of zeros. They operate on the components p,n of a quark, so that the transformations generated by F_1, F_2, F_3 are just rotations in isospin space. Thus $F_3 \equiv I_3$. There is one additional diagonal generator, F_8, related to the hypercharge Y. In fact, we can now derive the Gell-Mann-Nishijima relation. Assume

$$Q = aF_3 + bF_8 \qquad 5.26$$

and note that for a quark, the eigenvalues of F_8 are just

$$F_8 = \begin{pmatrix} \frac{1}{2\sqrt{3}} & & \\ & \frac{1}{2\sqrt{3}} & \\ & & -\frac{1}{\sqrt{3}} \end{pmatrix} \qquad 5.27$$

Now then, take for example $\pi^+ = p\bar{n}$. From the quantum number assignments to these components we have from 5.26

$$1 = a\left(\frac{1}{2} + \frac{1}{2}\right) + b\left(\frac{1}{2\sqrt{3}} - \frac{1}{2\sqrt{3}}\right) \qquad 5.28$$

Take now $K^+ = p\bar{\lambda}$

$$1 = a\left(\frac{1}{2} + 0\right) + b\left(\frac{1}{2\sqrt{3}} + \frac{1}{\sqrt{3}}\right) \qquad 5.29$$

Thus $a = 1,\ b = \frac{1}{\sqrt{3}}$,

$$Q = F_3 + F_8/\sqrt{3} \qquad 5.30$$

Comparing this with $Q = I_3 + \frac{Y}{2}$ we make the identification

$$F_3 = I_3; \quad F_8 = \frac{\sqrt{3}}{2}\, Y \qquad 5.31.$$

This is then the origin of the Y assignments to the quarks. By taking linear combinations of the generators we can obtain an alternative representation of SU(3), analogous to $\tau_\pm, \tau_3$ in SU(2). In fact this will emphasize that SU(3) is made out of 3 SU(2) subgroups, defining I-spin, U-spin, V-spin. We take

$$I_\pm = F_1 \pm iF_2; \quad I_+ = \begin{pmatrix} 0 & 1 & 0 \\ 0 & 0 & 0 \\ 0 & 0 & 0 \end{pmatrix}; \quad I_- = \begin{pmatrix} 0 & 0 & 0 \\ 1 & 0 & 0 \\ 0 & 0 & 0 \end{pmatrix}$$

$$U_\pm = F_6 \pm iF_7; \quad U_+ = \begin{pmatrix} 0 & 0 & 0 \\ 0 & 0 & 1 \\ 0 & 0 & 0 \end{pmatrix}; \quad U_- = \begin{pmatrix} 0 & 0 & 0 \\ 0 & 0 & 0 \\ 0 & 1 & 0 \end{pmatrix} \qquad 5.32$$

$$V^\pm = F_4 \mp iF_5; \quad V_+ = \begin{pmatrix} 0 & 0 & 0 \\ 0 & 0 & 0 \\ 1 & 0 & 0 \end{pmatrix}; \quad V_- = \begin{pmatrix} 0 & 0 & 1 \\ 0 & 0 & 0 \\ 0 & 0 & 0 \end{pmatrix}$$

Together with

$$I_3 = \begin{pmatrix} 1/2 & & \\ & -1/2 & \\ & & 0 \end{pmatrix}; \quad U_3 = \begin{pmatrix} 0 & & \\ & 1/2 & \\ & & -1/2 \end{pmatrix}; \quad V_3 = \begin{pmatrix} -1/2 & & \\ & 0 & \\ & & 1/2 \end{pmatrix}$$

as such alternative representation of SU(3).

Note that

$$I_3 + V_3 + U_3 = 0 \qquad 5.33$$

so that there are only 8 independent generators, as before. Note that there is a one to one correspondence between shift operators 5.32 and particle states or field operators for the octet. In fact

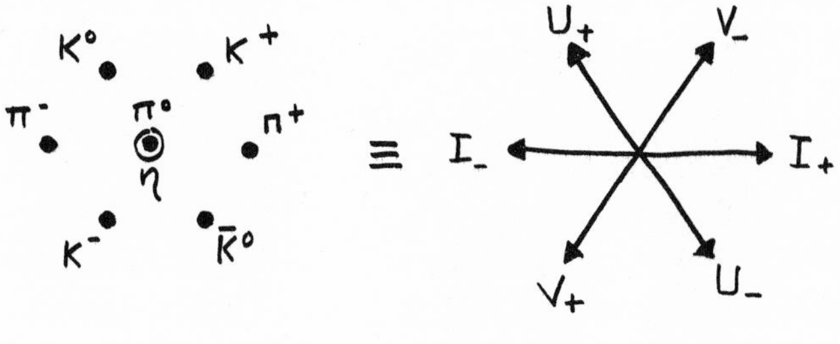

5.34

One could then write the scalar product 5.19 in terms of these new operators and obtain the analog of 4.17 for SU(2). The commutation relations of $I_\pm$, $V_\pm$, $U_\pm$ are of interest, they are just SU(2) commutation relations:

$$[I_3,I_\pm] = \pm I_\pm \qquad [I_3,U_\pm] = \pm\tfrac{1}{2}U_\pm \qquad [I_3,V_\pm] = \mp\tfrac{1}{2}V_\pm$$

$$[U_3,I_\pm] = \mp\tfrac{1}{2}I_\pm \qquad [U_3,U_\pm] = \pm U_\pm \qquad [U_3,V_\pm] = \mp\tfrac{1}{2}V_\pm$$

$$[V_3,I_\pm] = \mp\tfrac{1}{2}I_\pm \qquad [V_3,U_\pm] = \mp\tfrac{1}{2}U_\pm \qquad [V_3,V_\pm] = \pm V_\pm$$

$$[Y,I_\pm] = 0 \Longleftarrow \text{important} \qquad [I_+,I_3] = 2I_3 \qquad 5.35$$

$$[Y,U_\pm] = \pm U_\pm \qquad [U_+,U_-] = 2U_3$$

$$[Y,V_\pm] = \pm V_\pm \qquad [V_+,V_-] = 2V_3$$

and finally

$$[I_-,V_-] = U_+ \qquad [V_+,I_+] = U_-$$

$$[I_+,U_+] = V_- \qquad [U_-,I_-] = V_+ \qquad 5.36$$

$$[V^-,U_-] = I_+ \qquad [U_+,V_+] = I_-$$

<u>Product of two and three quarks</u>

As we shall see, the baryons will be constructed out of three quarks. It is easier to start with the intermediate step qq. We can do this with our graphical construction first:

qq :

$3 \otimes 3 = \bar{3} \oplus 6$

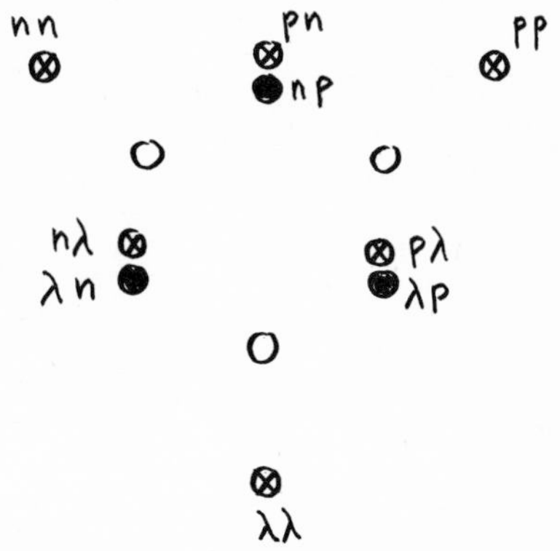

The construction of the linear combinations of the states appearing at the double points is particularly easy in this case. The upper row corresponds to the usual SU(2) $2 \otimes 2$ representation of isospin. The same can be said for the states appearing on the U-spin and V-spin axes. After taking symmetric and antisymmetric combinations of the states with mixed symmetry, we collect all symmetric states and find a <u>sextet</u>, while the antisymmetric states form a triplet

	I, I_3	Quark combination	
$\{\bar{3}\}$	0,0	$\frac{pn-np}{\sqrt{2}}$	
	$\frac{1}{2},\frac{1}{2}$	$\frac{p\lambda-\lambda p}{\sqrt{2}}$	5.37
	$\frac{1}{2},-\frac{1}{2}$	$\frac{(n\lambda-\lambda n)}{\sqrt{2}}$	

	I, I_3	Quark combination
{6}	1,1	pp
	1,0	$\frac{(pn+np)}{\sqrt{2}}$
	1,-1	nn
	$\frac{1}{2}, \frac{1}{2}$	$\frac{(p\lambda+\lambda p)}{\sqrt{2}}$
	$\frac{1}{2}, -\frac{1}{2}$	$\frac{(n\lambda+\lambda n)}{\sqrt{2}}$
	0	$\lambda\lambda$

5.38

From the tensor point of view we have

$$q_i q_j = T_{ij} = T_{|ij|} + T_{\{ij\}}$$

Now operate with ε^{abc}

$$\varepsilon^{ijk} T_{|ij|} = T^k \equiv q^k \equiv \bar{q}$$

3 components

Conversely

$\varepsilon^{ijk} T_{\{ij\}} = 0$, already irreducible;

$T_{\{ij\}}$ = Traceless, symmetric tensor with 9-3 = 6 components.

We can at this point give the general rules for the decomposition of a mixed tensor into a sum of irreducible representations. Consider the monomials

$$M(p,q) = q^\alpha q^\beta \ldots q^\gamma q_i q_j \ldots q_k = T^{\alpha\beta\ldots\delta}_{ij\ldots k} = T^{(p)}_{(q)} \qquad 5.39$$

with p upper and q lower indices. They transform accordingly to

$$T'^{\alpha,\beta\ldots\delta}_{i,j\ldots k} = U_{\alpha\lambda} U_{\beta\mu} \ldots U_{\delta\nu} U^*_{\ell i} U^*_{mj} \ldots U^*_{nk} T^{\lambda\mu\ldots\nu}_{\ell m\ldots n} \qquad 5.40$$

Irreducible representations will be those which cannot be reduced to lower rank or contracted by means of the tensors δ^i_j, ε^{ijk}, ε_{ijk}. We thus operate on $T^{\alpha\beta\ldots\delta}_{ij\ldots k}$ with these tensors and see under which condition this cannot be carried any further. We construct then

$$B^{\beta\ldots\delta}_{j\ldots\ell} = \delta^i_\alpha T^{\alpha\beta\ldots\delta}_{ij\ldots\ell} \qquad 5.41$$

This is just taking the trace of T.

$$C^{\gamma\ldots\delta}_{\mu ij\ldots\ell} = \varepsilon_{\mu\alpha\beta} T^{\alpha\beta\gamma\ldots\delta}_{ij\ldots\ \ell} \qquad 5.42$$

$$D^{m\alpha\beta\ldots\delta}_{k\ldots\ell} = \varepsilon^{mij} T^{\alpha\beta\ldots\delta}_{ijk\ldots\ell} \qquad 5.43$$

The tensor T is reducible unless B,C,D are identically zero.

$B = 0$ when $T^{i\beta\ldots\gamma}_{ij\ldots k} = 0$, namely when the trace with respect to α and i is zero.

$C = 0$ when T is symmetric in the indices α,β.

$D = 0$ when T is symmetric in the indices i,j.

To obtain irreducible representations of $SU(3)$ we must then take linear combinations $P(p,q)$ of the monomials $M(p,q)$ such that they are

1) Traceless
2) Totally symmetric in all p upper indices
3) Totally symmetric in all q lower indices.

} 5.44

The polynomials P(p,q) form a basis for the IR D(p,q) of SU(3). The dimension N of D(p,q) is

$$N = (1+p)(1+q)\left[1+\frac{1}{2}(p+q)\right] \qquad 5.45$$

For a proof of this and further details see, e.g., J. J. de Swart, Rev. Mod. Phys., 35, 916 (1963), also Y. A. Smorodinskii, Sov. Phys. USPEKHI, 84, 637 (1965). According to 5.45, the dimensionality of several representations are

$$\begin{array}{lll} D(0,0) = \{1\} & D(1,1) = \{8\} & D(0,3) = \{10\} \\ D(0,1) = \{3\} & D(0,2) = \{6\} & D(3,0) = \{\overline{10}\} \\ D(1,0) = \{\bar{3}\} & D(2,0) = \{\bar{6}\} & \text{etc.} \end{array} \qquad 5.46$$

We shall just sketch now the decomposition of the product of three quarks qqq.

$$3 \otimes 3 \otimes 3 = 1 \oplus 8 \oplus 8' \oplus 10 \qquad 5.47$$

We have just worked out $3 \otimes 3 = \bar{3} \oplus 6$, thus

$$3 \otimes 3 \otimes 3 = (\bar{3} + 6) \otimes 3 = \bar{3} \otimes 3 + 6 \otimes 3 \qquad 5.48$$

$\bar{3} \otimes 3 \to 1 \oplus 8$ (1: antis.; 8: mixed symmetry)

$6 \otimes 3 \to 8' \oplus 10$ (8': mixed symmetry; 10: symmetric)

The octets which appear here have mixed symmetry.

In fact the representation $\{\bar{3}\}$ in 5.48 is given by the antisymmetric combinations 5.37, so that the first octet is antisymmetric in two indices and symmetric in the third one. Conversely 8' is symmetric in two indices and antisymmetric in the third one. We could, of course, as we did before, obtain all the components of 1,8,8',10, in terms of the three-quark constituents. (See, e.g., B. Feld, Models of Elementary Particles, Blaisdell Publishing Co., Waltham, Mass., Toronto, London, 1969, p. 322). These however would not be too useful at this stage, since we would not know which of the two octets to identify with the N, Λ, Σ, Ξ baryon octet. In the next step to SU(6) where quarks are given two spin states, such ambiguity will disappear. For the moment we see that qqq is the most economical product representation to give singlets, octets and decuplets of baryons as observed. If any doubt should remain concerning the fact that baryon isospin multiplets can be made out of linear combinations of three quarks, note the following

Call $\left.\begin{matrix} p \\ n \end{matrix}\right\} = \nu \quad I = \frac{1}{2} \quad S = 0$

$\lambda = \lambda \quad I = 0 \quad S = -1$

Now, combinations involving

$\nu\nu\nu \rightarrow I = 1/2, 3/2$ will lead to $S = 0$ like N, Δ

$\nu\nu\lambda \rightarrow I = 0, 1,$ $S = -1$ like Λ, Σ

$\nu\lambda\lambda \rightarrow I = 1/2,$ $S = -2$ like Ξ

$\lambda\lambda\lambda \rightarrow I = 0,$ $S = -3$ like Ω^-.

It is important to realize that even if we do not have as yet an unambiguous assignment of the quark combinations making up the octet of stable baryons, the latter must have the same transformation properties under SU(3) transformations as, e.g., the octets of the mesons seen in detail before. Since there is only one "unequivalent" 8-dimensional representation of SU(3), we can represent the baryon octet in a way completely analogous to 5.15. By comparing the octet weight diagrams of mesons and baryons, we can write then

$$B_8(\tfrac{1}{2}^+) = \begin{pmatrix} \Sigma^0/\sqrt{2} + \Lambda/\sqrt{6} & \Sigma^+ & p \\ \Sigma^- & -\Sigma^0/\sqrt{2} + \Lambda/\sqrt{6} & n \\ \Xi^- & \Xi^0 & -2\Lambda/\sqrt{6} \end{pmatrix} \qquad 5.49$$

References

In addition to the references mentioned in the text, many sources have been consulted in the preparation of this lecture:

P.T. Matthews, Unitary Symmetry, in High Energy Physics, Vol. 1. Edited by E.H.S. Burhop, Academic Press, New York and London, 1967.

V.F. Weisskopf. $SU_2 \rightarrow SU_3 \rightarrow SU_6$. Lectures intended mainly for younger experimental physicists. CERN Report 66-19.

A. Morales. Higher Symmetries. Proceedings of the 1968 CERN School of Physics at El Escorial. CERN Report 68-23. Vol. I.

W.R. Frazer, Elementary Particles. Prentice Hall, Inc., 1966.

S.G. Gasiorowicz and S.L. Glashow. Unitary Symmetry, Advances in Theoretical Physics, Vol. II, Academic Press, 1966.

A particularly lucid series of lectures on SU(3), which the writer had the fortune to attend, was given by J. Prentki at CERN in 1964. Unfortunately the original notes for such lectures were lost. A similar course, incorporating some of Prentki's lecture material, was given in 1965 by:

L.C. Biedenharn, Group Theory and the Classification of the Elementary Particles. CERN Report 65-41.

The basic reference for a comprehensive coverage of the subject is the collection of classic papers and critical comments in:

M. Gell-Mann and Y. Ne'eman. The Eightfold Way. W.A. Benjamin, Inc., 1964.

6

SU(3) AS A BROKEN SYMMETRY

Some of the most interesting consequences of SU(3) arise from its being broken by several perturbations. If SU(3) were a rigorous symmetry for the strong interactions, all SU(3) multiplets should, e.g., be degenerate in mass. On the contrary, we note very large splittings amongst the various isospin multiplets composing an SU(3) multiplet, in addition to the small splitting among the members of isospin multiplets, attributed to the 1-spin breaking electromagnetic interaction. This is illustrated on page 48 for the meson and page 49 for the baryon lowest-lying multiplets.

Pseudoscalar meson octet: $J^P = 0^-$

Mass structure: SU(3) broken by MSI, EMI

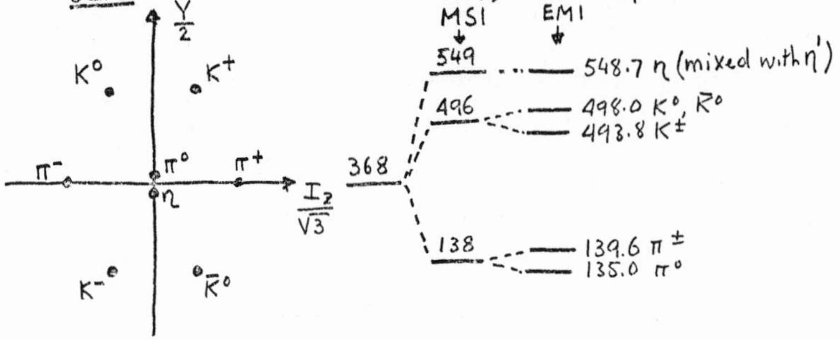

Pseudoscalar meson unitary singlet: $J^P = 0^-$
η' (958) (mixed with η)

Vector meson octet: $J^P = 1^-$

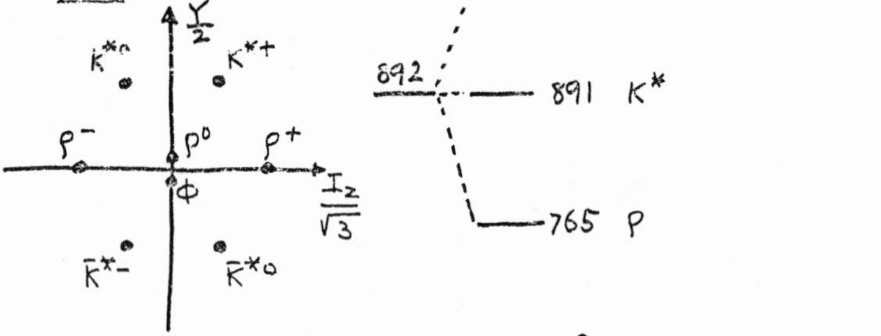

Vector meson unitary singlet: $J^P = 1^-$
ω(784) (mixed with ϕ)

Baryon octet: $J^P = \frac{1}{2}^+$

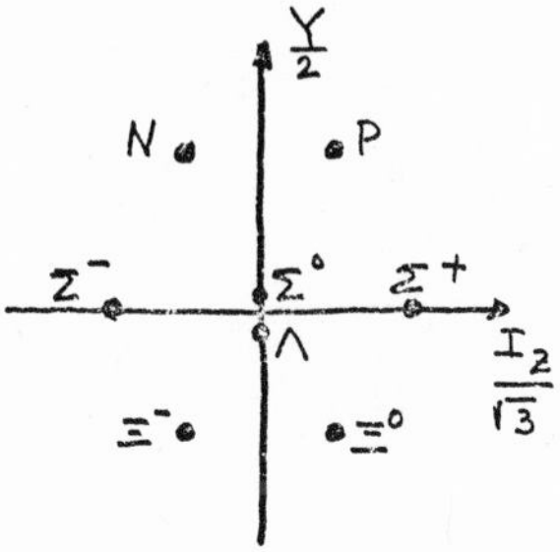

Mass structure:
SU(3) broken by:

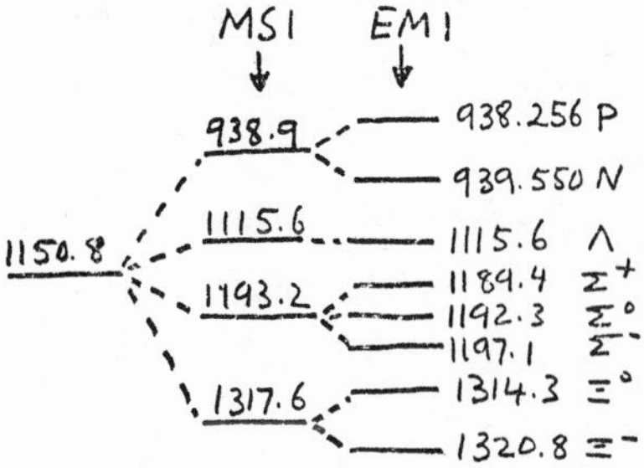

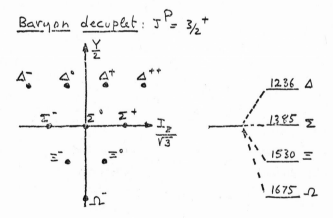

Baryon decuplet: $J^P = \frac{3}{2}^+$

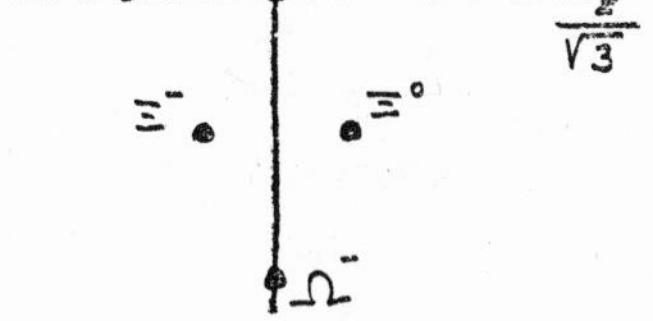

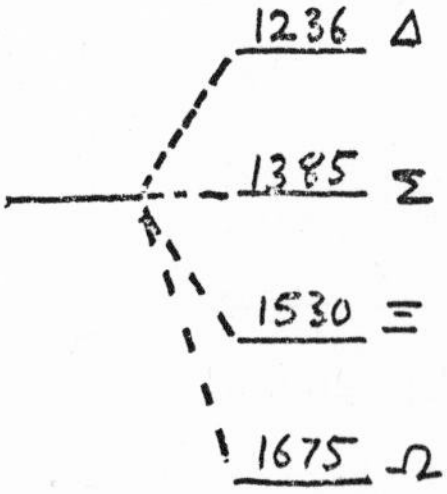

The Gell-Mann Okubo Mass Formula

Gell-Mann (see, e.g., M. Gell-Mann and Y. Ne'eman, The Eightfold Way, Benjamin, 1964) assumed that the large splittings observed in the meson and baryon multiplets arise as a result of a perturbation by a medium-strong interaction (MSI). Neglecting at this stage the smaller perturbation due to the E. M. interaction, since then the isospin multiplets are degenerate, the MSI must behave as a scalar in I-spin space. The symmetry breaking must then be attributed to either U-spin or V-spin non-conservation. (Note that the mass splitting is observed between members of, e.g., U-spin multiplets, like $P(U_3 = +\frac{1}{2})$ and $\Sigma^+(U_3 = -\frac{1}{2})$. If we choose to explore the behavior of the MSI along, say, the U-spin axis, it is convenient to rotate by -120° the usual octet Y-I_3 representation:

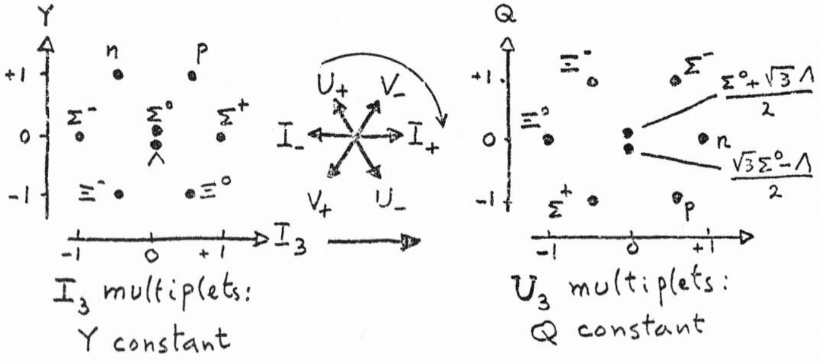

Since Σ^0, Λ are not eigenstates of U, they appear as such in a linear combination. This can be best seen by rotating $\begin{pmatrix} \Sigma^0 \\ \Lambda \end{pmatrix}$ by -120^0:

$$\begin{pmatrix} U_1^0 \\ U_0^0 \end{pmatrix} = - \begin{pmatrix} \cos\frac{2\pi}{3} & \sin\frac{2\pi}{3} \\ -\sin\frac{2\pi}{3} & \cos\frac{2\pi}{3} \end{pmatrix} \begin{pmatrix} \Sigma^0 \\ \Lambda \end{pmatrix} = \begin{pmatrix} \frac{1}{2} & -\frac{\sqrt{3}}{2} \\ \frac{\sqrt{3}}{2} & \frac{1}{2} \end{pmatrix} \begin{pmatrix} \Sigma^0 \\ \Lambda \end{pmatrix} \qquad 6.1$$

Note the effect of the $I_\pm, U_\pm$ operators once more. I_+, I_- change I_3 but do not change Y; U_+, U_- change Y, but not its electric charge Q. Thus U-spin multiplets are made of particles having the same coupling to the E.M. field. U-spin is conserved in the very strong interaction (VSI), assumed SU(3) invariant, but not conserved by the MSI. It is assumed that the MSI behaves as a scalar (S) in I-space, but as a vector (V) in U-space. In summary

<u>Interaction</u>	<u>I-spin</u>	<u>U-spin</u>	
VSI (SU(3) invariant)	S (conserved)	S (conserved)	
Electromagnetic	S + V	S (conserved)	6.2
MSI	S (conserved)	S + V	

Accordingly, we take the MSI mass operator in the form

$$O_M = M_0 + aU_3 \; ; \qquad O_M \Psi_i = M_i \Psi_i \qquad 6.3$$

with $M_0, a,$ scalars.

We now operate with O_M on the neutral U-spin triplet

$$\Xi^0, \; \frac{\Sigma^0 - \sqrt{3}\Lambda}{2}, \; n \; .$$

$$U_3 = -1, \quad 0, \quad +1$$

$$M_{\Xi^o} = \langle \Xi^o | O_M | \Xi^o \rangle = M_o - a \qquad 6.4$$

$$M_{U_1^o} = \left\langle \frac{\Sigma^o - \sqrt{3}\Lambda}{2} \middle| O_M \middle| \frac{\Sigma^o - \sqrt{3}\Lambda}{2} \right\rangle = M_o$$

$$= \frac{1}{4} \langle \Sigma^o | O_M | \Sigma^o \rangle + \frac{3}{4} \langle \Lambda | O_M | \Lambda \rangle = \frac{1}{4} M_{\Sigma^o} + \frac{3}{4} M_\Lambda \qquad 6.5$$

(Since O_M is I-spin conserving, there are no off diagonal elements between Λ, Σ^o.)

$$M_n = \langle n | O_M | n \rangle = M_o + a \qquad 6.6$$

Combining 6.4, 6.5, 6.6 we obtain the Gell-Mann Okubo sum-rule for the octet.

$$\boxed{2(N+\Xi) = 3\Lambda + \Sigma} \qquad 6.7$$

(Where the symbols stand for the particle masses.)

Consider now the $Q = -1$ U-spin quartet which appears in the baryon decuplet and apply 6.3:

$$\begin{array}{ccccc} & \Delta^- & \Sigma^- & \Xi^- & \Omega^- \\ U_3 & 3/2 & 1/2 & -1/2 & -3/2 \\ M = & M_o' + \frac{3}{2}a' & M_o' + \frac{1}{2}a' & M_o' - \frac{1}{2}a' & M_o' - \frac{3}{2}a' \end{array} \qquad 6.8$$

$$\Delta M = a' \qquad a' \qquad a'$$

which leads to the so-called equal spacing rule, i.e.,

$$\boxed{\Delta - \Sigma = \Sigma - \Xi = \Xi - \Omega} \qquad 6.8$$

Both 6.7 and 6.8 are special cases of the generalized Gell-Mann Okubo mass formula

$$M = a + bY + c\left[I(I+1) - Y^2/4\right] \qquad 6.9$$

(S. Okubo, Progr. Theoret. Phys. (Kyoto) 27, 949 (1962).

At the time of the proposal of the Eightfold Way by Gell-Mann, the possible existence of the representation {10} was predicted for the baryons as a consequence of the decomposition

$$8 \otimes 8 = 1 \oplus 8_1 \oplus 8_2 \oplus 10 \oplus \overline{10} \oplus 27 \qquad 6.10$$

since the members of the decuplet decay into mesons (8) and baryons (8). Possible candidates for the decuplet were however only the $\Delta(1236)$ and the $\Sigma(1385)$, both of $J^P = 3/2^+$, and isospin 3/2 and 1 respectively. As soon as the $\Xi(1530)$, $I = 1/2$ was discovered, its mass was noted to fit well the prediction of the equal spacing rule for the decuplet. The assignment of these three states to the {10} representation required unambiguously the existence of an isospin singlet, the Ω^-, whose mass could be predicted as 1676 MeV!

Such a particle, having strangeness $S = -3$, was discovered in 1964 at Brookhaven (Barnes et al., Phys. Rev. Letters, 12, 204 (1964)), after a worldwide hunt. The event, observed in the hydrogen bubble chamber, was initiated by 5 GeV/c K^- on protons:

$$K^- p \to \Omega^- + K^+ + K^0$$
$$\quad \hookrightarrow \Xi^0 + \pi^-$$
$$\qquad \hookrightarrow \Lambda + \pi^0$$
$$\qquad\quad \hookrightarrow \pi^- + p$$

The Ω^- mass from two events was found as (1675 ± 3)MeV!

Needless to say, after this spectacular triumph, SU(3) became considerably more popular than it had been before. Because of its mass, the Ω^- cannot decay strongly, conserving all the quantum numbers it should. It is thus the last of the "stable" baryons, decaying by weak interaction.

How well do the GMO sum rules fare today?

Mass, MeV (average for multiplet)

a) $\frac{1}{2}^+$ baryon octet	N	938.6	input
	Λ	1115.6	input
	Σ	1193.1	input
	Ξ	1331.4 predicted, 1318 observed	

b) $\frac{3}{2}^+$ decuplet	Δ	1236	input		
	Σ	1383	input		
The agreement is still	Ξ	1530	predicted	1531	observed
quite remarkable.	Ω	1677	predicted	1672.5	observed

The agreement is still quite remarkable.

For the pseudoscalar meson octet, 6.7 would read $2(K + \bar{K}) = 3\eta + \pi$, or, since

$$M_K = M_{\bar{K}}, \quad K = \frac{3\eta + \pi}{4} \qquad 6.11$$

This sum rule does not work so well. While as we have seen, the stable baryon octet satisfies 6.7 within $\sim 1\%$, 6.11 is correct only within $\sim 6\%$. If, however, 6.11 is rewritten for the <u>square</u> of the masses

$$\boxed{K^2 = \frac{3\eta^2 + \pi^2}{4}} \qquad 6.12$$

then the agreement becomes miraculously excellent again.

If we try to apply 6.7 to heavier baryon octets or 6.12 to heavier meson octets, we find trouble again. As we shall see, there are further consequences of symmetry breaking which will have to be taken into account.

Note that from the point of view of the quark model, the statement that $O_M = M_o + aU_3$ is equivalent to saying that the $\nu(p,n)$ quark has a mass which is different from the λ quark. If we assume that the λ quark is heavier than the ν quark, 6.7 and 6.8 follow at once.

$$\begin{array}{lll} N \equiv \nu\nu\nu & & \\ & \Delta M = \lambda - \nu & \\ \left.\begin{array}{l}\Sigma\\ \Lambda\end{array}\right\} \ \nu\nu\lambda & & 6.13 \\ & \Delta M = \lambda - \nu & \\ \Xi \equiv \nu\lambda\lambda & & \end{array} \qquad \begin{array}{l} \Xi^o - (\frac{1}{4}\Sigma^o + \frac{3}{4}\Lambda) = (\frac{1}{4}\Sigma^o + \frac{3}{4}\Lambda) - n \\ \frac{\Xi + n}{2} = \frac{3\Lambda + \Sigma}{4} \end{array}$$

Here we have reduced the octet to an equal spacing rule. 6.8 then becomes obvious as an extension of 6.13. The only trouble is that Σ should be as heavy as Λ!

Electromagnetic Mass Differences

The splittings of the members of an isospin multiplet are attributed to the isospin breaking electromagnetic interaction. On the other hand we have already remarked that all members of a U-spin multiplet have the same coupling to the e.m. field. Thus electromagnetic effects should be scalar in U-spin and, for example, the electromagnetic masses should be the same for all members of a U-spin multiplet. For the $\frac{1}{2}^+$ baryon octet this implies

$$\begin{aligned}
\langle p|O_E|p\rangle &= \langle \Sigma^+|O_E|\Sigma^+\rangle \rightarrow m_e(p) = m_e(\Sigma^+) \\
\langle \Sigma^-|O_E|\Sigma^-\rangle &= \langle \Xi^-|O_E|\Xi^-\rangle \rightarrow m_e(\Sigma^-) = m_e(\Xi^-) \\
\langle n|O_E|n\rangle &= \langle \Xi^0|O_E|\Xi^0\rangle \rightarrow m_e(n) = m_e(\Xi^o)
\end{aligned} \qquad 6.14$$

By setting $m_e(\Xi^-) - m_e(\Xi^0) = \Xi^- - \Xi^0$ = observable mass difference, we have from 6.14

$$\boxed{(\Xi^- - \Xi^0) = (\Sigma^- - \Sigma^+) - (n - p)}\,, \qquad 6.15$$

the Coleman-Glashow relation. (S. Coleman and S.L. Glashow, Phys. Rev. Letters 6, 423 (1961)).

Experimentally 6.15 gives

$$(6.5 \pm 0.2)\,\text{MeV} = (6.68 \pm 0.11)\,\text{MeV!}$$

Magnetic Moments of Baryons

The assumptions 6.2 for the properties of the e.m. interactions also explain the magnetic moments of baryons. This time we cast the magnetic moment operator μ in the form

$$\mu = \mu_o + aI_3 \qquad 6.16$$

Scalar plus I-spin vector. This gives immediately an "equal spacing rule" for the magnetic moments of the members of an isospin multiplet:

$$\mu_{\Sigma^+} - \mu_{\Sigma^o} = \mu_{\Sigma^o} - \mu_{\Sigma^-}$$

or

$$\mu_{\Sigma^+} + \mu_{\Sigma^-} = 2\mu_{\Sigma^o} \qquad 6.17$$

Also, from U-spin invariance:

$$\begin{aligned} \mu_p &= \mu_{\Sigma^+} \\ \mu_{\Sigma^-} &= \mu_{\Xi^-} \\ \mu_n &= \mu_{\Xi^o} = \mu_{U_1^o} \end{aligned} \qquad 6.18$$

In order to connect $\mu_{U_1^o}$ to physics, we perform some manipulations. First, from 6.1 we have

$$\begin{aligned} U_1^o &= \frac{\Sigma^o - \sqrt{3}\Lambda}{2} = A \\ U_o^o &= \frac{\sqrt{3}\Sigma^o + \Lambda}{2} = B \end{aligned} \rightarrow \begin{aligned} -\Lambda &= \frac{\sqrt{3}A - B}{2} \\ \Sigma^o &= \frac{A + \sqrt{3}B}{2} \end{aligned} \qquad 6.19$$

$$\begin{aligned} \mu_\Lambda &= \frac{\sqrt{3}A-B}{2}\left|\mu\right|\frac{\sqrt{3}A-B}{2} = \frac{3}{4}\mu_A + \frac{1}{4}\mu_B \\ \mu_{\Sigma^o} &= \frac{A+\sqrt{3}B}{2}\left|\mu\right|\frac{A+\sqrt{3}B}{2} = \frac{1}{4}\mu_A + \frac{3}{4}\mu_B \end{aligned} \qquad 6.20$$

$$2\mu_A = 3\mu_\Lambda - \mu_{\Sigma^o} = 2\mu_n \qquad 6.21$$

since $\mu_A = \mu_{U_1^o}$. Because of the symmetry of the octet, we also have $\sum_i \mu_i = 0$ or

$$\mu_{\Sigma^+} + \mu_{\Sigma^-} + \mu_{\Sigma^o} + \mu_\Lambda + \mu_p + \mu_n + \mu_{\Xi^-} + \mu_{\Xi^o} = 0 \qquad 6.22$$

From 6.22, 6.17, 6.18 we can write

$$2(\mu_{\Sigma^+} + \mu_{\Sigma^-}) + \mu_{\Sigma^0} + \mu_\Lambda + 2\mu_n = 0$$

or

$$5\mu_{\Sigma^0} + \mu_\Lambda + 2\mu_n = 0 \qquad 6.23$$

Using now 6.21 follows

$$\boxed{\mu_\Lambda = \frac{\mu_n}{2}} \qquad 6.24$$

$$\boxed{\mu_p + \mu_n = -\mu_{\Xi^-}} \qquad 6.25$$

The measured magnetic dipole moments so far are:

Baryon	Magnetic moment $(eh/2m_pc)$
p	$2.792782 \pm 1.7 \times 10^{-5}$
n	$-1.913148 \pm 6.6 \times 10^{-5}$
Λ	-0.73 ± 0.16 (from RPP
Σ^+	2.57 ± 0.52 August 1970)

As can be seen, the predictions which can be checked are reasonably well satisfied.

Mixing of SU(3) Representations

If we try to apply the GMO sum rule to the heavier octets, of both mesons and baryons, we run into difficulties. Take for example the vector mesons $J^P = 1^-$

$$\left.\begin{array}{l} K(\ 892) \\ \phi(1019) \\ \rho(\ 765) \end{array}\right] \text{—— input in GMO} \rightarrow \phi_{GMO}(928) \qquad 6.26$$

The experimental error on the masses are very small compared with this discrepancy. Similarly for the $\frac{3}{2}^-$ baryon octet:

$$\left.\begin{array}{l} N(1515) \\ \Lambda(1692) \\ \Sigma(1661) \\ \Xi(1819) \end{array}\right] \text{—— input in CMO} \rightarrow \Lambda_{GMO}(1669) \qquad 6.27$$

Note that in both examples (and there are many more), there is a unitary singlet having the same quantum numbers as the $I=0, I_3=0$ member of the octet.

<u>ω(784)</u>	$J^P = 1^-$,	{1}	<u>Λ(1520)</u>	$J^P = \frac{3}{2}^-$
<u>ϕ(1019)</u>	$J^P = 1^-$,	{8}	<u>Λ(1692)</u>	$J^P = \frac{3}{2}^-$

A possible explanation for the observed discrepancies between the GMO predictions and the observed values of the masses can be found if it is assumed that the MSI, which breaks SU(3), can lead to <u>mixing</u> between members of different unitary multiplets having the same quantum numbers. The observed physical states would then not be pure SU(3) states, but a mixture, in this case, of singlet and octet states.

A mixing angle θ can be defined:

$$\begin{aligned} \langle 8| &= \langle \Lambda_8|\cos\theta + \langle \Lambda_1|\sin\theta \\ \langle 1| &= -\langle \Lambda_8|\sin\theta + \langle \Lambda_1|\cos\theta \end{aligned} \qquad 6.28$$

(left side: pure SU(3) states; right side: physical states)

The mixing angle can be derived from the observed masses as follows:

Write the Hamiltonian for pure SU(3) states

$$H = H_o + V = \begin{pmatrix} m_o(8)+v_{88} & v_{18} \\ v_{81} & m_o(1)+v_{11} \end{pmatrix} \equiv \begin{pmatrix} m_8 & m_{18} \\ m_{81} & m_1 \end{pmatrix} \qquad 6.29$$

m is a mass matrix operator, linear in the masses for baryons, linear in the $(\text{masses})^2$ for the mesons. Thus

$$H\begin{pmatrix} \psi_8 \\ \psi_1 \end{pmatrix} = \begin{pmatrix} m_8 & m_{18} \\ m_{81} & m_{11} \end{pmatrix}\begin{pmatrix} \psi_8 \\ \psi_1 \end{pmatrix} \; ; \qquad H\psi = m\psi \qquad 6.30$$

The observed particles are eigenstates of H:

$$H' \begin{pmatrix} \psi'_8 \\ \psi'_1 \end{pmatrix} = \begin{pmatrix} m_{8obs} & 0 \\ 0 & m_{1obs} \end{pmatrix} \begin{pmatrix} \psi'_8 \\ \psi'_1 \end{pmatrix}; \quad H'\psi' = m'\psi' \qquad 6.31$$

The similarity transformation

$$H' = MHM^{-1} \qquad 6.32$$

reduces m to diagonal form, where $\psi' = M\psi$, or

$$\begin{pmatrix} \psi'_8 \\ \psi'_1 \end{pmatrix} = \begin{pmatrix} \cos\theta & -\sin\theta \\ \sin\theta & \cos\theta \end{pmatrix} \begin{pmatrix} \psi_8 \\ \psi_1 \end{pmatrix} \qquad 6.33$$

From 6.32 and 6.33 we obtain

$$\sin^2\theta = \frac{m_{8,obs} - m_8}{m_{8,obs} - m_{1,obs}} \qquad \sin^2\theta = \frac{m^2_{8,obs} - m^2_8}{m^2_{8,obs} - m^2_{1,obs}}$$

for the baryons, for the mesons. 6.34

Of course we do not know m_8. We assume however that m_8 is given correctly by the GMO relation. Using the observed masses in 6.34 we have, for example, the following situations

$0^-; \theta = 10.4^o$ $1^-; \theta = 40.1^0$ $\frac{3}{2}^-; \theta = 21.4^o$

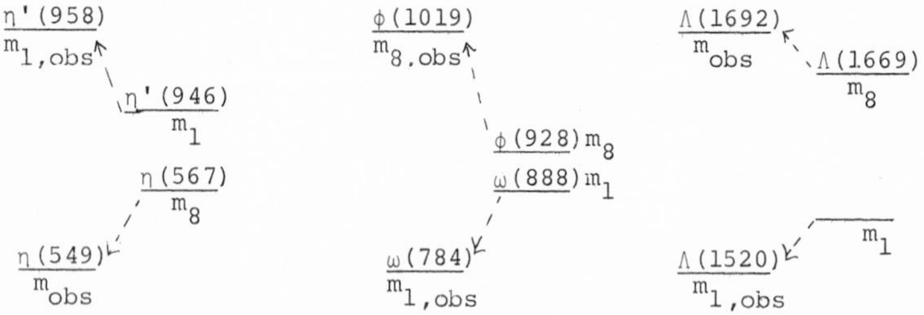

The pseudoscalar meson octet is relatively pure. The hypothesis of mixing between SU(3) representations is more than just a recipe to save the validity of the GMO relation.

We shall see that the mixing angle can be derived independently of the GMO formula, from the observed decay rates of baryons and mesons.

A particularly convincing test is provided by the following example. Consider the decay

$$\begin{array}{ccc} \Lambda(1520) & \rightarrow & \Sigma(1385)\pi \\ \{1\} & & \{10\} \times \{8\} \end{array} \qquad 6.35$$

This decay is forbidden by unbroken SU(3), since

$$\{10\} \times \{8\} = \{8\} + \{10\} + \{27\} + \{35\} \qquad 6.36$$

There is no coupling between the singlet and the 10×8 representation. This is an example of an SU(3) selection rule. On the other hand the decay 6.34 would be allowed for the $3^-/2$ member of the octet

$$\Lambda(1692) \rightarrow \Sigma(1385)\pi \qquad 6.37$$

The decay 6.35 has indeed been observed to occur at a substantial rate.

References

In several derivations it was found most profitable to follow the approach illustrated by H.J. Lipkin. Unitary Symmetry for Pedestrians. Argonne National Laboratory Informal Report, August 1963. See also C.A. Levinson, J.H. Lipkin and S. Meshkov, Phys. Rev. Letters, 10, 561 (1963).

B.T. Feld. Models of Elementary Particles. Blaisdell Publ. Co., 1969, and The Quark Model of the Elementary Particles. CERN Report 67-21.

In addition, all references to Lecture 5 are of course relevant to the content of Lecture 6.

7
PARTICLE REACTIONS AND STRONG DECAY PROCESSES

SU(3) invariance gives definite predictions about coupling strengths in particle reactions, branching ratios in particle decay, in analogy to SU(2) invariance which involves specific predictions about cross sections and decay rates based on isospin conservation.

Simple predictions can be derived by exploiting U-spin conservation. This reduces the SU(3) problem to simple SU(2). (See References to Lecture 6.) Consider for example the reactions

$$\begin{aligned} \pi^- + p &\to K^+ + \Sigma^- \\ &\to \pi^+ + \Delta^- \\ K^- + p &\to K^+ + \Xi^- \\ &\to \pi^+ + \Sigma^- \end{aligned}$$

$$U:\quad \underbrace{\tfrac{1}{2}\qquad \tfrac{1}{2}}\qquad \underbrace{\tfrac{1}{2}\qquad \tfrac{3}{2}}$$

$$U_{tot}:\quad 0,1 \qquad\qquad 1,2$$

only channel (between the 1 of "0,1" and the 1 of "1,2")

Where $\Sigma^-, \Delta^-, \Xi^-$ are the members of the $3/2^+$ decuplet.

First of all we see that U-spin conservation requires all these reactions to proceed through the $U = 1$ channel.

All reactions can then be described in terms of a single amplitude a_1.

The combination coefficients in this case are the product of two SU(2) Clebsch-Gordan coefficients:

$$\langle U_1 m_1 U_2 m_2 | U_1' m_1' U_2' m_2' \rangle = (U_1 m_1 U_2 m_2 | UM)(U_1' m_1' U_2' m_2' | UM)\, a_1$$

$$\langle \pi^- p | K^+ \Sigma^- \rangle = (\tfrac{1}{2}\ \tfrac{1}{2}\ \tfrac{1}{2}\ \tfrac{1}{2} | 11)(\tfrac{1}{2}\ \tfrac{1}{2}\ \tfrac{3}{2}\ \tfrac{1}{2} | 11)\, a_1 = -\tfrac{1}{2}\, a_1$$

$$\langle \pi^- p | \pi^+ \Delta^- \rangle = (\tfrac{1}{2}\ \tfrac{1}{2}\ \tfrac{1}{2}\ \tfrac{1}{2} | 11)(\tfrac{1}{2}\ {-\tfrac{1}{2}}\ \tfrac{3}{2}\ \tfrac{3}{2} | 11)\, a_1 = \tfrac{\sqrt{3}}{2}\, a_1$$

$$\langle K^- p | K^+ \Xi^- \rangle = (\tfrac{1}{2}\ {-\tfrac{1}{2}}\ \tfrac{1}{2}\ \tfrac{1}{2} | 10)(\tfrac{1}{2}\ \tfrac{1}{2}\ \tfrac{3}{2}\ {-\tfrac{1}{2}} | 10)\, a_1 = -\tfrac{1}{2}\, a_1$$

$$\langle K^- p | \pi^+ \Sigma^- \rangle = (\tfrac{1}{2}\ {-\tfrac{1}{2}}\ \tfrac{1}{2}\ \tfrac{1}{2} | 10)(\tfrac{1}{2}\ {-\tfrac{1}{2}}\ \tfrac{3}{2}\ \tfrac{1}{2} | 10)\, a_1 = \tfrac{1}{2}\, a_1$$

The prediction is then (apart from kinematical corrections) that the four reactions should proceed at relative rates in the ratio 1: 3: 1: 1 respectively. Similar consideration can be applied to determine relative decay rates. (See homework problem.) Such an approach bypasses the use of SU(3) Clebsch-Gordan coefficients.

The derivation of the SU(3) Clebsch-Gordan coefficients can be undertaken at various levels of sophistication. In their simplest form they have been tacitly derived as the coefficients appearing in 5.15 for e.g., the coupling of $\{3\} \times \{\bar{3}\}$ to $\{8\}$ and $\{1\}$, or in 5.37 and 5.38 for the coupling of $\{3\} \times \{3\}$ to $\{\bar{3}\}$ and $\{6\}$.

A two-particle reaction

$$A + B \rightarrow A' + B' \qquad 7.1$$

in which each component is a member of a definite SU(3) multiplet, can be viewed in the limit of exact SU(3) symmetry, as a two-step process in which an intermediate state is formed, also a member of a definite SU(3) multiplet.

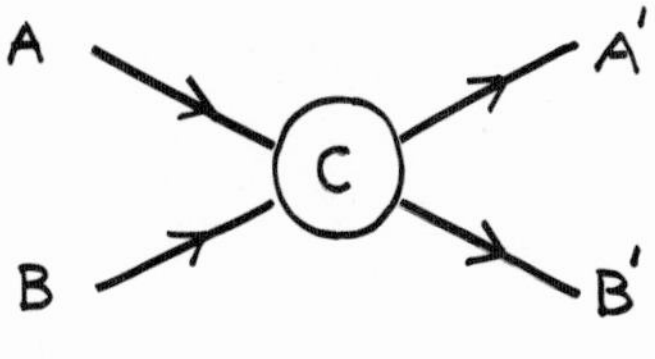

7.1

In fact, the transition amplitude for the process 7.1 will be, in SU(3)

$$T(AB \rightarrow A'B') = \langle A' \otimes B' | T | A \otimes B \rangle \qquad 7.2$$

where $|A \otimes B\rangle = |A\rangle \otimes |B\rangle = |\phi_A, \rho_A\rangle \otimes |\phi_B, \rho_B\rangle$ is the product representation of two SU(3) states characterized by a representation $\{\phi_A\}$ and $\{\phi_B\}$ and a suitably chosen set of quantum numbers ρ_A and ρ_B respectively. The same for $|A' \otimes B'\rangle$. We know that the product $|A \otimes B\rangle$ can be reduced to a sum of irreducible representations of the <u>Clebsch-Gordan series</u>

$$\{\phi_A\} \otimes \{\phi_B\} = \Sigma \sigma_C \{\phi\}_C \qquad 7.3$$

as seen in many examples. Here σ_C is an integer to indicate how many times the same D(p,q) appears in the series.

The "projection" of $|A \otimes B\rangle$ on the irreducible representations of the C.G. series is

$$|A \otimes B\rangle = \sum_{\phi_C, \gamma, \rho_C} \begin{pmatrix} \phi_A \phi_B \phi_{C,\gamma} \\ \rho_A \rho_B \rho_C \end{pmatrix} |\phi_{C,\gamma}, \rho_C\rangle \qquad 7.4$$

where the parentheses are the SU(3) Clebsch-Gordan coefficients. This is simply a more formal expression of what we have done, e.g., in 5.37 and 5.38. Clearly 7.4 gives the relative coupling strengths of particles A,B, to the intermediate state C. A similar relation will exist for $|A' \times B'\rangle$. Then

$$T(AB \rightarrow A'B') = \sum_{\substack{\phi_C, \gamma, \rho_C \\ \phi_i \gamma' \rho_{C'}}} \begin{pmatrix} \phi_{A'} \phi_{B'} \phi_{C'\gamma'} \\ \rho_{A'} \rho_{B'} \rho_{C'} \end{pmatrix} \begin{pmatrix} \phi_A \phi_B \phi_{C\gamma} \\ \rho_A \rho_B \rho_C \end{pmatrix} \langle \phi_{ij} \rho_i | T | \phi_{C,\gamma} \rho_C \rangle \qquad 7.5$$

For exact SU(3) symmetry, we want T to be a scalar, and the only non-zero matrix elements $\langle \phi_{C'\gamma'} \rho_{C'} | T | \phi_{C,\gamma} \rho_C \rangle$ are those for which $\rho_C = \rho_{C'}, \phi_C = \phi_{C'}$. This is of course the origin of the SU(3) selection rules, as mentioned on page 61. Both initial and final states in 6.36 must couple to the same intermediate state, or only the common channels through the representation $\phi_C = \phi_{C'}$ are open. 7.5 then reduces to

$$T(AB \rightarrow A'B') = \sum_{\substack{\phi_C \rho_C \\ \gamma\ \gamma'}} \begin{pmatrix} \phi_{A'} \phi_{B'} \phi_{C'\gamma'} \\ \rho_{A'} \rho_{B'} \rho_{C'} \end{pmatrix} \begin{pmatrix} \phi_A \phi_B \phi_{C\gamma} \\ \rho_A \rho_B \rho_C \end{pmatrix} \underbrace{T(\phi_{C,\gamma})}_{\substack{\text{Invariant} \\ \text{SU(3) matrix} \\ \text{elements.}}} \qquad 7.6$$

The case of a strong decay $C \rightarrow A + B$ 7.7

is already contained as a subcase of the above, the decay amplitudes being determined by 7.4.

The specification of the basis vectors $|A\rangle, |B\rangle, |C\rangle$ has been so far only symbolic. As discussed by de Swart, RMP 35, 916 (1963), a convenient choice for the basic vectors is the eigenvector of a set of commuting operators, such as

$$|C\rangle = |\{\phi_C\}, \{\phi_A\}, \{\phi_B\}, I^2, I^2(A), I^2(B), I_3, Y\rangle \qquad 7.8$$

where $\{\phi_C\}$ characterizes one of the I.R. of the direct product $\{\phi_A\} \otimes \{\phi_B\}$,

$$I^2 = [I(A) + I(B)]^2; \quad I_3 = I_3(A) + I_3(B), \; Y = Y(A) + Y(B) \qquad 7.9$$

In terms of this basis, 7.4 can be reformulated as

$$|\{\phi_A\}, I^2(A), I_3(A), Y(A)\rangle \otimes |\{\phi_B\}, I^2(B), I_3(B), Y(B)\rangle = \sum_{\{\phi_C\}, I, Y} C^{I(A)I(B)I(C)}_{I_3(A)I_3(B)I_3(C)} \begin{pmatrix} \{\phi_A\} & \{\phi_B\} & \{\phi_C\} \\ I(A)Y(A) & I(B)Y(B) & I\ Y \end{pmatrix} \times \qquad 7.10$$

$$|\{\phi_C\}, \{\phi_A\}, \{\phi_B\} I^2, I_3, Y\rangle$$

where, following de Swart, the SU(3) Clebsch-Gordan coefficients have been represented as the product of SU(2) C.G. coefficients <u>isoscalar factors</u>.

$$\begin{pmatrix} \phi_A & \phi_B & \phi_C \\ \rho_A & \rho_B & \rho_C \end{pmatrix} = C^{I(A)I(B)I(C)}_{I_3(A)I_3(B)I_3(C)} \begin{pmatrix} \{\phi_A\} & \{\phi_B\} & \{\phi_C\} \\ I(A)Y(A) & I(B)Y(B) & I\ Y \end{pmatrix} \qquad 7.11$$

SU(2) C.G. coefficients × isoscalar factors

The SU(2) C.G. coefficients are well known, the isoscalar factors are tabulated by de Swart.

We now have the apparatus essential to predict relative strengths in particle reactions and decays. Since several of the interesting couplings involve the <u>product representation</u> of <u>two octets</u>, we still have to elaborate on a particular complication which arises in this case. Consider for example an octet representation, like 5.49

$$B^i_k = \begin{pmatrix} \Sigma^0/\sqrt{2} + \Lambda/\sqrt{6} & \Sigma^+ & p \\ \Sigma^- & -\Sigma^0/\sqrt{2} + \Lambda/\sqrt{6} & n \\ \Xi^- & \Xi^0 & -2\Lambda/\sqrt{6} \end{pmatrix} \qquad 7.12$$

where the symbols are only a short-hand for indicating the components explicitly. As an extension of 5.8 we can construct a traceless tensor with 8 components out of the product of two octets as follows

$$O^\ell_k = B^i_k B'^\ell_i - \frac{1}{3}\begin{pmatrix} I & 0 & 0 \\ 0 & I & 0 \\ 0 & 0 & I \end{pmatrix} \qquad 7.13$$

where I is the invariant

$$I = B^i_k B'^k_i = \delta^\ell_k B^i_k B'^\ell_i = \Sigma^o\Sigma'^o + \Lambda\Lambda' + \Sigma^+\Sigma'^+ + \Sigma^-\Sigma'^- \qquad 7.14$$
$$+ \Xi^-\Xi'^- + \Xi^o\Xi'^o + nn' + pp'$$

Consider now a specific term of O^ℓ_k:

$$O^2_1 = B^1_1 B'^2_1 + B^2_1 B'^2_2 + B^3_1 B'^2_3 = \qquad 7.15$$

$$\frac{1}{\sqrt{6}}\Lambda\Sigma'^+ + \frac{1}{\sqrt{2}}\Sigma^0\Sigma'^+ + \frac{1}{\sqrt{6}}\Sigma^+\Lambda' - \frac{1}{\sqrt{2}}\Sigma^+\Sigma'^0 + p\Xi'^o$$

O^2_1 is antisymmetric in the $\Sigma\Sigma$ terms, symmetric in the $\Lambda\Sigma$ terms, and has mixed symmetry in the $p\Xi'^0$ term. One could on the other hand construct a different octet:

$$\tilde{O}^\ell_k = B^k_i B'^i_\ell - \frac{1}{3}\begin{pmatrix} I & 0 & 0 \\ 0 & I & 0 \\ 0 & 0 & I \end{pmatrix} \qquad 7.16$$

The $\tilde{O}^2_1$ component would read now

$$\tilde{O}^2_1 = B^2_1 B'^1_1 + B^2_2 B'^2_1 + B^2_3 B'^3_1 = \qquad 7.17$$

$$\frac{1}{\sqrt{6}}\,\Sigma^+\Lambda' + \frac{1}{\sqrt{2}}\,\Sigma^+\Sigma'^0 + \frac{1}{\sqrt{6}}\,\Lambda\Sigma'^+ - \frac{1}{\sqrt{2}}\,\Sigma^0\Sigma'^+ + \Xi^0 p'$$

Octets of definite symmetry can be constructed by taking proper linear combinations of 7.13 and 7.16. It is customary to define, for example

$$A^k_\ell = \tfrac{1}{2}(O^k_\ell - \tilde{O}^k_\ell) = \underline{\text{antisymmetric octet}}, \quad \{8_a\}$$

$$S^k_\ell = \tfrac{1}{2}(O^k_\ell + \tilde{O}^k_\ell) = \underline{\text{symmetric octet}}, \quad \{8_s\} \qquad 7.18$$

This is then the origin of the two octets in

$$8 \otimes 8 = 1 \oplus 8_a \oplus 8_s \oplus 10 \oplus \overline{10} \oplus 27. \qquad 7.19$$

To connect with 7.4, 7.19 is a case in which the sum over the subscript γ has to be performed for $\gamma = 1,2$. As a result, the coupling of two octets to give another octet leads to <u>two</u> invariant amplitudes. Schematically then

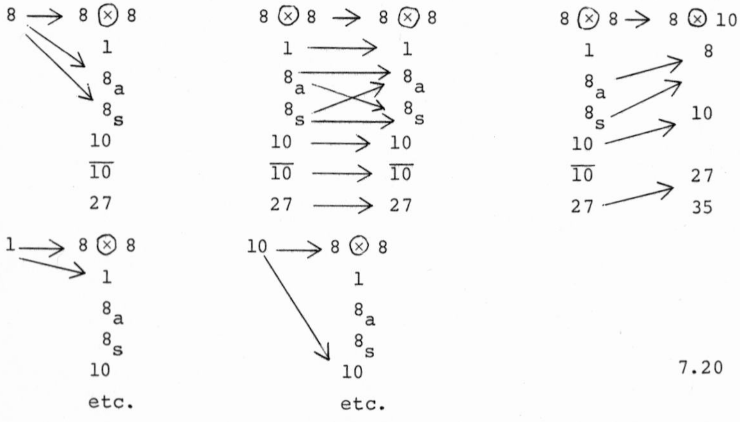

7.20

where each arrow corresponds to a non-zero term in the sum 7.6 or 7.4. In practice not all the transitions allowed by SU(3) are actually occurring due to either the presence of selection rules arising from other symmetries, or due to particular dynamical features of the interaction, e.g., scattering process involving the formation of an intermediate resonant state.

Yukawa Coupling in SU(3)

In analogy with the SU(2) case, seen previously (see 4.23, 4.24) we want to construct a scalar to describe the interaction vertex $\bar{B}BM$, where each of the particles belongs to octet representations. As we have just seen in 7.18, there are two ways to combine two octets to give another octet; in our case 7.18 will read

$$\begin{aligned} &\text{antisymmetric octet} \qquad A = \tfrac{1}{2}(B\bar{B} - \bar{B}B) \\ &\text{symmetric octet} \qquad S = \tfrac{1}{2}(B\bar{B} + \bar{B}B) \end{aligned} \qquad 7.21$$

We now have to make a scalar out of the products AM and SM. This is accomplished by taking the traces. This then defines two types of coupling

$$\begin{aligned} &\text{symmetric coupling} \qquad \underline{\text{D type}}: \ \mathrm{Tr}[(B\bar{B} + \bar{B}B)M] \\ &\text{antisymmetric coupling} \qquad \underline{\text{F type}}: \ \mathrm{Tr}[(B\bar{B} - \bar{B}B)M] \end{aligned} \qquad 7.22$$

There are now in the literature various ways to define such couplings and the confusion can become overwhelming.

We can for example write the Hamiltonian in the form

$$H = \sqrt{2}\, D\{Tr(\bar{B}MB) + Tr(\bar{B}BM)\} + \sqrt{2}\, F\{Tr(\bar{B}MB) - Tr(\bar{B}BM)\} \qquad 7.23$$

The coupling constants D and F relate to the notation of de Swart as follows:

$$D = \frac{\sqrt{30}}{40}\, g_1 \qquad\qquad g_1 = g_D$$

$$F = \frac{\sqrt{6}}{24}\, g_2 \qquad 7.24 \qquad \text{Also found are} \qquad g_2 = g_F \qquad 7.25$$

and $g = D + F$

$$\alpha = \frac{F}{F+D} \qquad 7.26 \qquad\qquad \text{or} \quad \alpha = \frac{D}{D+F} \qquad 7.27$$

(de Swart) (Gell-Mann)

From 7.23, with

$$B = \begin{pmatrix} \Sigma^0/\sqrt{2} + \Lambda\ \sqrt{6} & \Sigma^+ & p \\ \Sigma^- & -\Sigma^0/\sqrt{2} + \Lambda/\sqrt{6} & n \\ \Xi^- & \Xi^0 & -2\Lambda/\sqrt{6} \end{pmatrix};$$

$$\bar{B} = \begin{pmatrix} \bar{\Sigma}^0/\sqrt{2} + \bar{\Lambda}/\sqrt{6} & \bar{\Sigma}^- & \bar{\Xi}^- \\ \bar{\Sigma}^+ & -\bar{\Sigma}^0/\sqrt{2} + \bar{\Sigma}/\sqrt{6} & \bar{\Xi}^0 \\ \bar{p} & \bar{n} & -2\bar{\Lambda}/\sqrt{6} \end{pmatrix}; \qquad 7.28$$

$$M = \begin{pmatrix} \pi^0/\sqrt{2} + n/\sqrt{6} & \pi^+ & K^+ \\ \pi^- & -\pi^0/\sqrt{2} + n/\sqrt{6} & K^0 \\ K^- & \bar{K}^0 & -2n/\sqrt{6} \end{pmatrix},$$

we obtain

$$H = (D+F)\bar{N}\vec{\tau}N\vec{\pi} + (\sqrt{3}F-\frac{1}{\sqrt{3}}D)\bar{N}N\eta-(\sqrt{3}F+\frac{1}{\sqrt{3}}D)(\bar{N}K\Lambda+\bar{\Lambda}\bar{K}N) +$$

$$(D-F)(\bar{\vec{\Sigma}}\bar{K}\vec{\tau}N+\bar{N}\vec{\tau}K\vec{\Sigma}) + 2iF\bar{\vec{\Sigma}}\times\vec{\Sigma}\vec{\pi} + \frac{2D}{\sqrt{3}}\bar{\vec{\Sigma}}\vec{\Sigma}\eta +$$

$$\frac{2}{\sqrt{3}}D(\bar{\Lambda}\vec{\Sigma}+\bar{\vec{\Sigma}}\Lambda)\vec{\pi} - \frac{2}{\sqrt{3}}D\bar{\Lambda}\Lambda\eta + (F-D)\bar{\Xi}\vec{\tau}\Xi\vec{\pi} - (\sqrt{3}F+\frac{D}{\sqrt{3}})\bar{\Xi}\Xi\eta +$$

$$\frac{2}{\sqrt{3}}D(\bar{\Xi}K_C\Lambda+\bar{\Lambda}\bar{K}_C\Xi) - (D+F)(\bar{\vec{\Sigma}}\bar{K}_C\vec{\tau}\Xi+\bar{\Xi}\vec{\tau}K_C\vec{\Sigma}) \qquad 7.29$$

$$\text{where}\quad N = \begin{pmatrix} p \\ n \end{pmatrix},\quad K = \begin{pmatrix} K^+ \\ K^0 \end{pmatrix},\quad K_C = \begin{pmatrix} \bar{K}_0 \\ -K^- \end{pmatrix},\quad \Xi = \begin{pmatrix} -\Xi^0 \\ \Xi^- \end{pmatrix}, \qquad 7.30$$

$\vec{\Sigma}, \vec{\pi}$ are isotriplets, Λ, η isoscalars.

The table on page 72 collects the couplings of practical interest for the transitions

$\{1\} \to \{8\} \times \{8\}$, $\{8\} \to \{8\} \times \{8\}$, $\{10\} \to \{8\} \times \{8\}$, e.g.,

$\Lambda(1520) \to \Sigma + \pi$ $\Sigma(1760) \to \Lambda + \pi$ $\Delta(1236) \to N + \pi$

The couplings are expressed as the products

$$g_{\bar{B}MB} = (\text{isoscalar factor}) \times (\text{reduced SU(3) matrix element})$$
$$(I_1Y_1I_2Y_2|IY) \qquad \times (\text{e.g., } g_1,\ g_D,\ g_F,\ g_{10}) \qquad 7.31$$

These still have to be multiplied by the appropriate SU(2) C.G. coefficients, if reactions involving specific charge states are considered.

Remarks: 1) The entries for the Yukawa coupling $\{8\}_{D,F}$ are nothing else than the coefficients of the corresponding terms of 7.29 as can be easily checked.

2) The same form of the $\{8\}_{D,F}$ coupling holds for any three octets. Thus we could use the table also for the coupling between three meson octets. <u>However</u> we must take

into account in this case of a selection rule which stems effectively from the requirement of Bose statistics for the mesons. Under the operation of charge conjugation, the pseudoscalar mesons transform as

$CMC^{-1} = \tilde{M}$ while the vector mesons as $CVC^{-1} = -\tilde{V}$ (like the photon) 7.32

Because of C-invariance then, in the decay

$V \to M + M$ e.g., $\rho \to \pi + \pi$	7.33	the symmetric D-coupling vanishes and only the F coupling contributes
$V \to V + M$	7.34	the reverse is true and only D coupling contributes.

In practice, if the products of the charge conjugation q.n. C of the neutral members of the three meson octets is positive → D coupling, if negative → F coupling.

Coupling constant $g_{\bar{B}MB}$ for $\bar{B}$ member of {1}, $\{8\}_{D,F}$ and {10}.

		{1}	{8}D_1F	{10}
$N^*_{\frac{1}{2}}$	$\to N\pi$		$\frac{3}{2\sqrt{5}}g_D + \frac{1}{2}g_F$	
(N)	ΣK		$-\frac{3}{2\sqrt{5}}g_D + \frac{1}{2}g_F$	
	$N\eta$		$-\frac{1}{2\sqrt{5}}g_D + \frac{1}{2}g_F$	
	ΛK		$-\frac{1}{2\sqrt{5}}g_D - \frac{1}{2}g_F$	
$N^*_{\frac{3}{2}}$	$\to N\pi$			$-\frac{1}{\sqrt{2}}g_{10}$
(Δ)	ΣK			$\frac{1}{\sqrt{2}}g_{10}$

Coupling constant $g_{\bar{B}MB}$ for $\bar{B}$ member of {1}, $\{8\}_{D,F}$ and {10} (continued).

		{1}	{8}D_1F	{10}
$Y_0^* \to$	$N\bar{K}$	$\frac{1}{2}g_1$	$\frac{1}{\sqrt{10}}g_D + \frac{1}{\sqrt{2}}g_F$	
(Λ)	ΞK	$-\frac{1}{2}g_1$	$-\frac{1}{\sqrt{10}}g_D + \frac{1}{\sqrt{2}}g_F$	
	$\Sigma\pi$	$\frac{\sqrt{3}}{\sqrt{8}}g_1$	$-\frac{\sqrt{3}}{\sqrt{5}}g_D$	
	$\Lambda\eta$	$-\frac{1}{\sqrt{8}}g_1$	$-\frac{1}{\sqrt{5}}g_D$	
$Y_1^* \to$	$\bar{N}\bar{K}$		$-\frac{\sqrt{3}}{\sqrt{10}}g_D + \frac{1}{\sqrt{6}}g_F$	$-\frac{1}{\sqrt{6}}g_{10}$
(Σ)	ΞK		$-\frac{\sqrt{3}}{\sqrt{10}}g_D - \frac{1}{\sqrt{6}}g_F$	$\frac{1}{\sqrt{6}}g_{10}$
	$\Sigma\pi$		$\frac{\sqrt{2}}{\sqrt{3}}g_F$	$\frac{1}{\sqrt{6}}g_{10}$
	$\Sigma\eta$		$\frac{1}{\sqrt{5}}g_D$	$\frac{1}{2}g_{10}$
	$\Lambda\pi$		$\frac{1}{\sqrt{5}}g_D$	$-\frac{1}{2}g_{10}$
$\Xi^* \to$	$\Xi\pi$		$-\frac{3}{2\sqrt{5}}g_D + \frac{1}{2}g_F$	$\frac{1}{2}g_{10}$
(Ξ)	$\Sigma\bar{K}$		$\frac{3}{2\sqrt{5}}g_D + \frac{1}{2}g_F$	$\frac{1}{2}g_{10}$
	$\Xi\eta$		$-\frac{1}{2\sqrt{5}}g_D + \frac{1}{2}g_F$	$\frac{1}{2}g_{10}$
	$\Lambda\bar{K}$		$-\frac{1}{2\sqrt{5}}g_D + \frac{1}{2}g_F$	$-\frac{1}{2}g_{10}$

References

A recent comparison of SU(3) predictions with experiment for particle reactions is given by G.H. Trilling, Tests of SU(3) in particle reactions. Nuclear Physics B40 (1972), 13.

In discussing the Clebsch-Gordan Series, the references which have been consulted are:

J.J. de Swart. The Octet model and its Clebsch-Gordan coefficients. Rev. Mod. Physics, 35, 916 (1963).

A. Morales. Higher Symmetries. Proceedings of the 1968 CERN School of Physics at El Escorial, CERN Report 68-23, Vol. 1.

G. Costa, Introduction to Unitary Symmetries. Proceedings 1964 Herceg-Novi Easter School. CERN Report 64-13.

The problem of two octets has been discussed following V.F. Weisskopf. $SU_2 \rightarrow SU_3 \rightarrow SU_6$. CERN Report 66-19.

8

SU(3) CLASSIFICATION OF MESONS AND BARYONS

We have now reviewed most of the experimental consequences of SU(3) invariance, as well as the consequences of symmetry breaking. Beyond the point of establishing SU(3) as a valid symmetry, even if only approximate in some cases, the predictions of SU(3) provide the most valuable tool available for a coherent classification of particle states. Turning the emphasis around, from what has been introduced up to here, we can summarize the conditions for the classification of a set of particles in a given SU(3) representation. Such particles should

a) All have the same spin-parity J^P.

b) All have hypercharge and isospin properties contained in the representation. In addition:

c) The number of particle should equal, or at most have one extra component, the dimensionality of the representation.

d) The particle masses should satisfy the Gell-Mann-Okubo mass formula for the representation, taking into account the possibility of mixing.

e) The decay rates for the strong decays of each particle should satisfy the predictions based on the appropriate SU(3) coupling. (If mixing is present, the strength of such mixing should be consistent with d).)

f) The reaction amplitudes involving such particles should exhibit the appropriate phase relation when compared with those of other SU(3) representations.

We have dealt already with items a,b,c,d in some detail. Concerning e) and f) we need some further specification.

Decay Widths and SU(3) Coupling Constants

What determines the decay widths is not only SU(3). We must take into account the kinematical factors. For this purpose different recipes have been used, based on some theoretical arguments or other. For the decay of the baryons, for example, a formula based on non-relativistic potential theory has been used: (R.D. Tripp et al., Nuclear Phys. B3, 10 (1967).

$$\text{For } \{1\},\ \{10\} \qquad \Gamma = (cg)^2 B_\ell(K) \frac{M_N}{M_R}\, k$$

$$\{8\} \qquad \Gamma = (c_D g_D + c_F g_F)^2 B_\ell(k) \frac{M_N}{M_R}\, k \qquad 8.1$$

Here Γ is a partial decay width, c and g are respectively the SU(3) Clebsch-Gordan coefficient and the coupling constant g_1 or g_{10} and similarly for the octet which relates to the two coupling constants g_D and g_F, as in the table at page 72. k is the center of mass channel momentum. $B_\ell(k)$ is a barrier penetration factor (see, e.g., Blatt and Weisskopf).

$$B_\ell(k) = \frac{(kr)^{2\ell}}{D_\ell}, \qquad D_\ell = \begin{cases} 1 & \ell = 0 \\ 1 + (kr)^2 & \text{for } \ell = 1 \\ 9 + 3(kr)^2 + (kr)^4 & \ell = 2 \\ \text{---} & \text{---} \end{cases} \qquad 8.2$$

r is the radius of interaction taken as

$$1 \text{ Fermi} = \frac{1}{0.1973} \text{ GeV}^{-1}.$$

For mesons instead, a similar relation where the barrier factor is supposed to be relativistically invariant has been used (M. Goldberg, ANL Symposium on the present status of SU(3) for particle couplings and reactions, July 1967). Using the present notation:

$$\Gamma = (cg)^2 \frac{k}{M_R^2} \left(\frac{k^2 x^2}{k^2 + x^2} \right)^{\ell} \qquad \text{with} \quad x = \frac{1}{r} \qquad 8.3$$

In principle, for those decays in which only one coupling constant is involved, like {1} and {10}, one well measured rate determines g and then all other rates can be predicted and compared with experiment. If also the radius of interaction is left as a free parameter, then we need at least two measured rates. In practice one can for example minimize the χ^2 function using all the rates available for a given multiplet.

$$\chi^2 = \Sigma \frac{(\Gamma_{SU(3)} - \Gamma_{exp})^2}{(\Delta\Gamma_{exp})^2} \qquad 8.4$$

As mentioned previously, there will be only one coupling constant also for certain meson decays which couple $\{8\} \rightarrow \{8\} \times \{8\}$. For example vector meson → pseudoscalar + pseudoscalar, when only g_F is retained. The coupling constants are affected by singlet-octet mixing as was the case for the masses, as seen previously. In particular if we set

$$G_1 = c_1 g_1 \qquad \text{for} \quad \{1\} \to \{8\} \times \{8\}$$

$$G_8 = c_D g_D + c_F g_F \qquad \{8\} \to \{8\} \times \{8\} \tag{8.5}$$

$$\begin{cases} G_8 = G_8' \cos\theta - G_1' \sin\theta \\ G_1 = G_8' \sin\theta + G_1' \cos\theta \end{cases} \tag{8.6}$$

where G_8', G_1' represent the physical coupling constants, derived from 8.1.

When both g_D and g_F are involved, these can either be determined having been left as free parameters in 8.4, or more explicitly, they can be obtained from the following method, derived by R.D. Tripp.

Determination of octet coupling constants

From, e.g., 8.1, each decay rate defines two straight lines

$$c_D g_D + c_F g_F = \pm \left[\frac{\Gamma}{B_\ell(k) \frac{M_N}{M_R} k} \right]^{1/2} \tag{8.7}$$

in the g_D - g_F plane, whose slopes are determined by the ratio of the SU(3) isoscalar factors. The distance from the origin to the intercept with either axis is proportional to $\sqrt{\Gamma}$. If a particular channel becomes uncoupled, its line will pass through the origin. One looks for a common intercept of all "channel lines" which are available for a given octet,

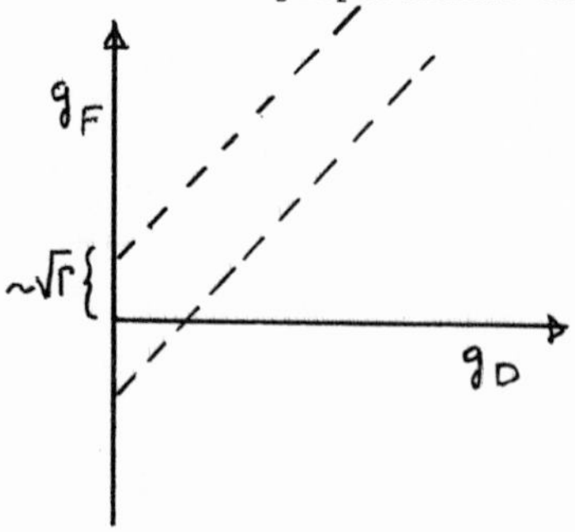

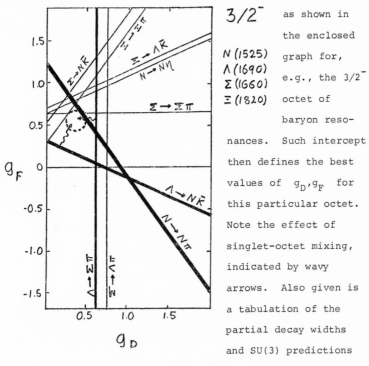

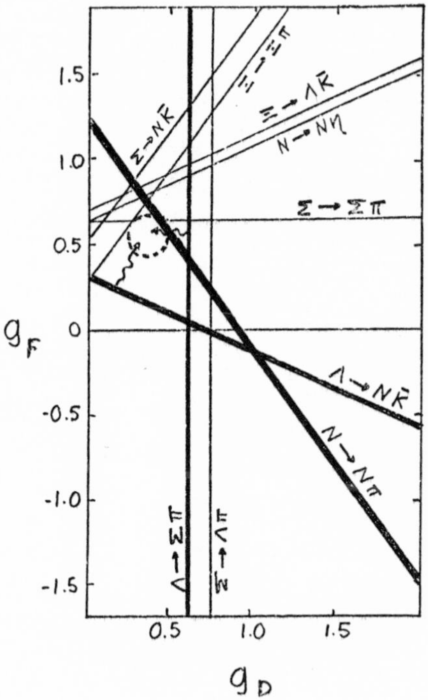

as shown in the enclosed graph for, e.g., the $3/2^-$ octet of baryon resonances. Such intercept then defines the best values of g_D, g_F for this particular octet. Note the effect of singlet-octet mixing, indicated by wavy arrows. Also given is a tabulation of the partial decay widths and SU(3) predictions for the same multiplet, as a practical example of the

Partial decay widths for baryon states assigned to the $3/2^-$ octet. Arrows indicate the consequence of singlet-octet mixing of Λ(1520) with Λ(1690).

Mass	Width	Mode	Branching fraction	Partial width	$B_\ell(k)k\frac{M_N}{M_R}$	$\left(\frac{\Gamma}{B_\ell(k)k\frac{M_N}{M_R}}\right)^{\frac{1}{2}}$	{8} Partial width	SU(3) sign of resonant amplitude
N(1525)	115	Nπ	0.55	57	154	0.608	38	+
		Nη	∿0.005	∿ 0.6	3.88	0.393	0.16	+
Λ(1690)	40	NK̄	0.18	7.2	117	0.248	28 →11.7	+
		Σπ	0.60	24	101	0.487	7 →19.2	-
		Λη	<0.005	< 0.2	1.81	0.332	.042→.047	-
Σ(1660)	50	NK̄	0.09	4.5	100	0.212	0.11	+
		Λπ	0.29	14.5	127	0.338	2.9	+
		Σπ	0.49	24.5	87.0	0.531	17	+
Ξ(1820)	20	Ξπ	<0.10	< 2.0	97.4	0.143	0.17	
		ΛK̄	0.30	6.0	86.7	0.263	3.5	
		ΣK̄	0.30	6.0	39.3	0.391	9.7	

procedure. What one actually calculates are the "multiplet widths," obtained from 8.1 using as input the best values, in this case, of g_D and g_F. The multiplet widths can then be compared with the observed widths. This is shown for a group of established multiplets in a graph taken from Levi Setti (1969), where the ratio of the two widths is plotted on a logarithmic scale. The scatter of the points away from the value 1.0 is a measure of how well, or how incompletely, SU(3) can account for observations which would otherwise bear no relation whatsoever to each other.

Ratio of the Multiplet Width to Observed Width for Baryon Resonances

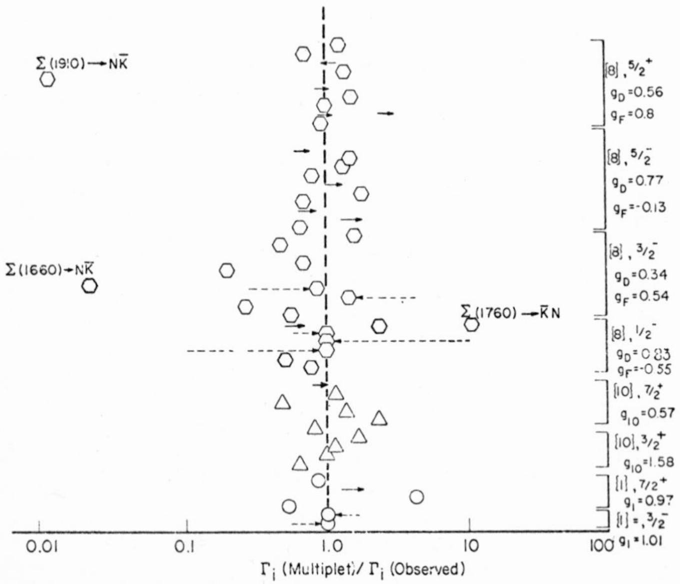

Relative Sign of Resonant Amplitudes and SU(3) Assignment

Once the coupling constants have been determined, the SU(3) Clebsch-Gordan coefficients require that specific phase relations be satisfied between the resonant amplitudes of particles belonging to different SU(3) representations, provided the SU(3) assignments have been made correctly. In fact in a formation-type experiment, or S-channel

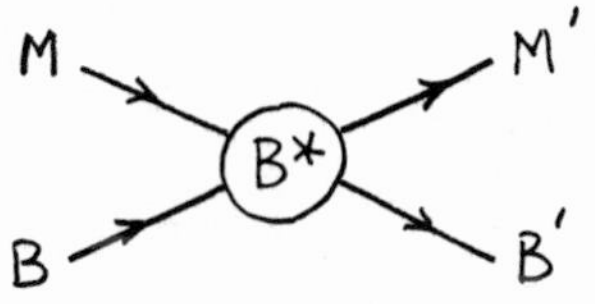

reaction, the resonant amplitude T_{res} can be be written as

$$T_{res} \sim \frac{g_{BMB^*} \cdot g_{B'M'B^*}}{E_R - E - i\Gamma/2} \qquad 8.8$$

with $\phi = \pm 1$. The coupling constants are those given at page 72. The absolute sign of one resonant amplitude is of course undetermined by an arbitrary phase. However, after having chosen a particular phase convention, then the relative sign of two resonant amplitudes becomes a meaningful SU(3) prediction. Empirically, the sign of a resonant amplitude is determined by means of a phase shift analysis of the reaction, and is measured as + or - in relation to the behaviour of the resonating partial wave in the Argand diagram. Here again the enclosed plot gives a survey of the relative

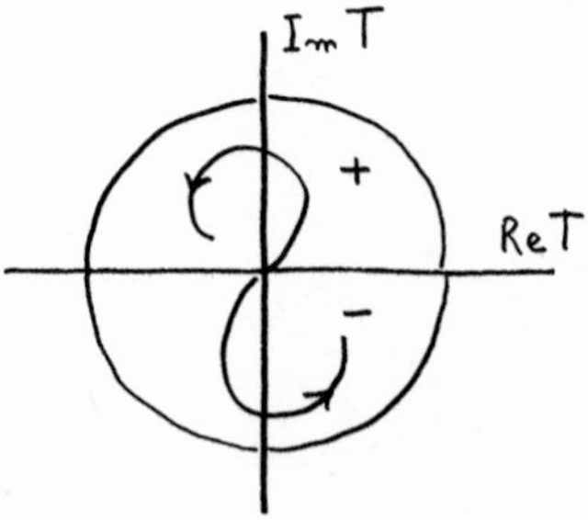

signs so far established compared with the SU(3) prediction. This is then both a test of the SU(3) assignment for baryons, for example, and at the same time, of the relevance of SU(3) is describing baryon classification.

SU$_3$ RELATIVE SIGN OF RESONANT AMPLITUDES

$T_{RES} \sim \phi\,(g_{N\bar{K}Y^*} \cdot g_{Y\pi Y^*})/(E_R - E - i\,\Gamma/2)$

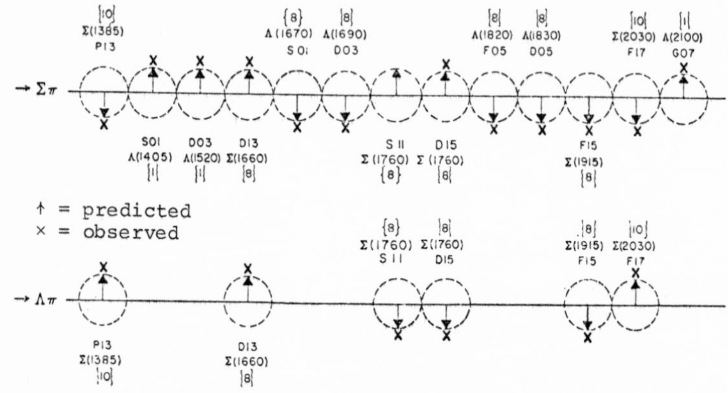

<u>Singlet-octet Mixing and Decay Rates</u>

Will describe just one relevant example. We have already seen that the deviation from the Gell-Mann-Okubo mass formula for the $3/2^-$ baryon octet could be accounted for in terms of singlet-octet mixing, for $|\theta| = \underline{(21.4 \pm 4)^0}$

______ Λ(1690), {8}

______ Λ(1520), {1}

Two recent experiments, one at LRL of the formation type, one by CERN-Chicago-Heidelberg-Saclay, in which Λ(1520) was studied as a final state, in $K^- p \to \Lambda(1520)\pi^0$, have yielded accurate values for the branching fractions:

$$\Lambda(1520) \to \bar{K}N(45.6 \pm 1.1)\%$$
$$\to \Sigma\pi(41.0 \pm 1.1)\%.$$

Using the value Γ = (15.7±1.2)MeV for the total width of Λ(1520) and the Λ(1690) partial widths for the corresponding channel, contained in the tabulation given, from Eq. 8.6 one obtains $\underline{\theta = -(18.3\pm2)^{\circ}}$, $g_D = 0.40 \pm 0.07$, $g_F = 0.42 \pm 0.06$, $g_1 = 0.99 \pm 0.05$. The mixing angle obtained in this way is quite consistent with that derived from the GMO relation. The hypothesis then that the apparent symmetry breaking effects are to be attributed to configuration mixing, appears well substantiated. Even more so if one considers the evidence for the decay $\Lambda(1520) \rightarrow \Sigma(1385)\pi$, mentioned previously. The actual rate of decay in this process however has recently been found much larger than expected under the hypothesis of singlet-octet mixing alone, for $\theta \approx 20^{\circ}$ (T. S. Mast et al., 1971). A clarification of this difficulty in terms of mixing with an additional octet predicted by the quark model has been reported (D. Faiman and D. E. Plane, 1972).

References

In addition to the references mentioned in the text, the material in this lecture is discussed by, among others,

R. D. Tripp, Proceedings of the 14th International Conference on High Energy Physics, Vienna (1968) (CERN, Geneva).

R. Levi Setti, Proceedings of the 1969 Lund International Conference on Elementary Particles (Institute of Physics, Lund, Sweden).

Recent developments are contained in papers by

D. E. Plane et al., Nuclear Physics B22, 93(1970).

T. S. Mast et al., Phys. Rev. Letters 28, 1220(1972).

D. Faiman and D. E. Plane, Phys. Lett. 39B, 358(1972).

9

QUARK MODELS AND SU(6)

In these lectures, we have been talking about quarks for the very introduction of SU(3). We have in fact looked at SU(3) from the most primitive form of quark model. The quarks however have been just useful tools to construct SU(3) representations and to obtain a better grasp of the more formal aspects of SU(3). Under the generic name of quark model however, a model (in fact several versions) is usually meant in which several kinds of dynamical assumptions are made about the interactions among quarks. Conjectures are made about the forces which bind quarks together, and these are described in terms of various kinds of potentials. Since the quarks are attributed spin $\frac{1}{2}$, central forces may be considered ($\vec{\sigma}\cdot\vec{\sigma}$), or non-central forces ($\vec{\sigma}\cdot\vec{L}$ or tensor) to act between quarks. The quarks become more and more physical in their attributes, to the point that their existence is seriously anticipated and evidence searched for. We can only undertake here, however, just to mention some of the developments of this memorable story. In 1964, Gursey and Radicati, and Sakita, proposed the group SU(6) as the symmetry which would go a step further than SU(3) in explaining the structure and properties of the particle spectrum. A given representation of SU(6) can be decomposed into a sum of SU(3) multiplets of dimension N, each carrying spin value j (representation of SU(2)) of dimension (2j+1). This will be indicated by $\{N_{2j+1}\}$. By assigning spin $\frac{1}{2}$ to the quarks, $\sigma = \frac{1}{2}$, it is clear that one can obtain spin

0^- mesons from $q\bar{q}$ in the $^1S_0$ state 

1^- mesons from $q\bar{q}$ in the $^3S_1$ state 

$\frac{1}{2}$ baryons from qqq up-up-down 

$\frac{3}{2}$ baryons from qqq up-up-up 

The basic representations of SU(6) have dimension 6, namely three SU(3) states with spin up, and three SU(3) states with spin down. These are the quarks of Gell-Mann and Zweig.

$$q_i = \begin{pmatrix} p^\uparrow \\ n^\uparrow \\ \lambda^\uparrow \\ p^\downarrow \\ n^\downarrow \\ \lambda^\downarrow \end{pmatrix}; \quad \bar{q}^i = (\bar{p}^\uparrow \bar{n}^\uparrow \bar{\lambda}^\uparrow \bar{p}^\downarrow \bar{n}^\downarrow \bar{\lambda}^\downarrow) \quad \text{or } p_1 \text{ for } p^\uparrow, \; p_2 \text{ for } p^\downarrow \text{ etc.} \qquad 9.1$$

By following the same procedure of 5.7, 5.8, etc., we would find for $q\bar{q}$

$$q \otimes \bar{q} = 6 \otimes \bar{6} = 1 \oplus 35 \qquad \begin{matrix} 1 = \text{symmetric in } \sigma, I \\ 35 = \text{mixed symmetry} \end{matrix} \qquad 9.2$$

In terms of SU(3) and spin components,

$$\begin{matrix} 35 = 8_1(j=0) \oplus 1_3(j=1) + 8_3(j=1) \\ 8 \quad + 3 \quad + 24 = 35 \end{matrix} \qquad 9.3$$

This is the first phenomenal result: The lowest meson multiplets <u>fit in</u> and <u>fill</u> these representations!

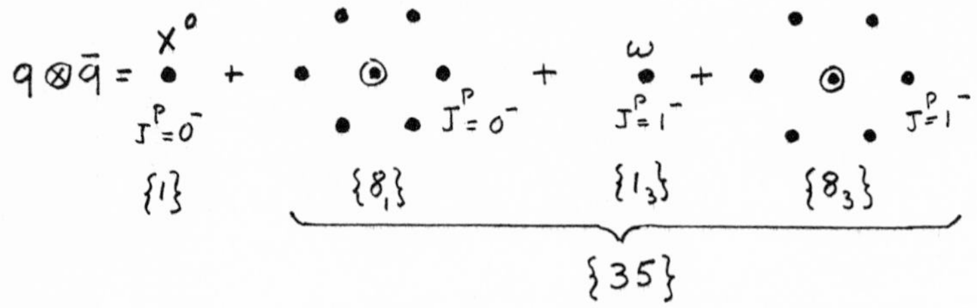

Remember the large mixing in the vector meson nonet of SU(3) and the small mixing of the pseudoscalar meson nonet of SU(3). From 9.2 and 9.3 we see that in SU(6) the singlet and octet pseudoscalar mesons belong to <u>different</u> representations, {1} and {35}, while all the 1^- mesons belong with {35}. As the SU(3) MSI leads to the mass splitting within the octet, it becomes natural to attribute to the SU(6) mass-breaking effects, also the mixing of the vector mesons. We can take now a different approach to the problem of mixing of the vector mesons as follows: We have already seen that symmetry breaking is in the direction of giving the <u>λ quark</u> a mass heavier than that of the degenerate <u>ν quark</u>. Guess what this mass difference amounts to. Take two simple states, m = 1

$p^+ = p^{\uparrow}\bar{n}^{\uparrow} = 765$ MeV $\quad \Delta m = m_\lambda - m_\nu \sim 127$ MeV

$K^{*+} = p^{\uparrow}\bar{\lambda}^{\uparrow} = 892$ MeV Consider now the mass difference:

$\phi - \omega = (1020 - 784)$ MeV $= 236$ MeV. Not far from $2\Delta m$.

We could guess that the observed ϕ is made out of λ quarks, the observed ω made out of ν-quarks, for the $m = 1$ components,

$$\phi_{obs} = \lambda^{\uparrow}\bar{\lambda}^{\uparrow}$$

$$\omega_{obs} = \frac{1}{\sqrt{2}} (p^{\uparrow}\bar{p}^{\uparrow} + n^{\uparrow}\bar{n}^{\uparrow}) \qquad 9.4$$

Now, what would be the relation between these states and the SU(6) states (the analogs of 5.10 and 5.13, taking the spin into account)?

$$\begin{pmatrix} \frac{1}{\sqrt{2}}(p^{\uparrow}\bar{p}^{\uparrow}+n^{\uparrow}\bar{n}^{\uparrow}) \\ \lambda^{\uparrow}\bar{\lambda}^{\uparrow} \end{pmatrix} = \begin{pmatrix} \sqrt{2/3} & \sqrt{1/3} \\ \sqrt{1/3} & -\sqrt{2/3} \end{pmatrix} \begin{pmatrix} \frac{1}{\sqrt{3}}(p^{\uparrow}\bar{p}^{\uparrow}+n^{\uparrow}\bar{n}^{\uparrow}+\lambda^{\uparrow}\bar{\lambda}^{\uparrow}) \\ \frac{1}{\sqrt{6}}(p^{\uparrow}\bar{p}^{\uparrow}+n^{\uparrow}\bar{n}^{\uparrow}- 2\lambda^{\uparrow}\bar{\lambda}^{\uparrow}) \end{pmatrix} \qquad 9.5$$

or

$$\begin{pmatrix} \omega_{obs} \\ \phi_{obs} \end{pmatrix} = \begin{pmatrix} \sqrt{2/3} & \sqrt{1/3} \\ \sqrt{1/3} & -\sqrt{2/3} \end{pmatrix} \begin{pmatrix} \phi_1 \\ \omega_8 \end{pmatrix} \qquad 9.6$$

Is this anything like 6.32? In fact, it is, since

$$\begin{pmatrix} \cos\theta & \sin\theta \\ \sin\theta & -\cos\theta \end{pmatrix} \begin{pmatrix} \phi_1 \\ \omega_8 \end{pmatrix} \equiv \begin{pmatrix} \cos\theta & -\sin\theta \\ \sin\theta & \cos\theta \end{pmatrix} \begin{pmatrix} \omega_8 \\ \phi_1 \end{pmatrix}$$

$$\theta \to \theta - \frac{\pi}{2}$$

It then follows that $\sin\theta = \sqrt{1/3} = 0.577$ or $\underline{\theta = (35.3)}^{\circ}$, which is then a prediction of the quark model. This is not far from the value $\theta \sim 40^{\circ}$ which follows from use of the GMO relation and the empirical masses.

One could now, following any of the methods indicated for SU(3) construct all the meson states in terms of their quark components, introduce the generators of SU(6), etc. We will just mention a few more consequences. In SU(6) the product of three quarks becomes

$$q \otimes q \otimes q = 6 \otimes 6 \otimes 6 = 20 \oplus 56 \oplus 70 \oplus 70' \qquad 9.7$$

The states making up 20 are antisymmetric, those making up 56 are symmetric, and 70 and 70' have mixed symmetry.

The representation 56 contains

$$56 = 8_2 \oplus 10_4 \quad . \qquad 9.8$$

or just an octet of spin 1/2 and a decuplet of spin 3/2!

The ambiguity of 8 and 8' in SU(3), coming from qqq has disappeared. There is now only one symmetric octet (which is in fact a linear combination of the two). The lowest baryon states fit in and fill the {56} representation of SU(6)! Nature seems to have chosen to populate only this representation, which is totally symmetric, for the baryon ground states. (Note the problems with Fermi statistics, however). The other possibilities in fact would contain

$$\begin{aligned} 20 &= 1_4 \oplus 8_2 \\ 70 &= 1_2 \oplus 8_2 \oplus 8_4 \oplus 10_2 \end{aligned} \qquad 9.9$$

which do not correspond to what we see. The symmetric {56} corresponds to {10} of SU(3). Number of symmetric states out of 3 quarks, each with n basic states:

$$N = \frac{n(n+1)(n+2)}{6} \qquad \begin{aligned} n = 3 &\rightarrow N = 10 \\ n = 6 &\rightarrow N = 56 \end{aligned} \qquad 9.10$$

For the complete wave functions of mesons and baryon states in SU(6), see, e.g., B. T. Feld, p. 327. Among the successes of SU(6), we should mention the correct prediction of the ratio of proton to neutron magnetic moments

$$\frac{\mu_n}{\mu_p} = -\frac{2}{3} \qquad 9.11$$

a mass formula which gives the masses of the baryon octet and decuplet in terms of 3 constants, and similarly for the mesons, and actual predictions of decay rates, without free parameters, and many others.

We should still mention the extension of the quark model to include orbital angular momentum.

In the system $q\bar{q}$, the mass splitting due to different angular momentum states is given by a centrifugal term of the form

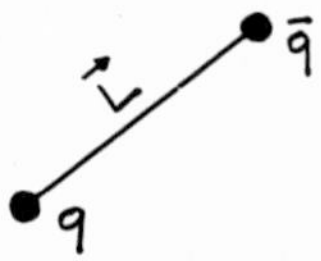

$$\delta(M^2) = L(L+1)\left\langle\frac{1}{r^2}\right\rangle_L \qquad 9.11$$

Since L is known to be linear in M^2 (Chew-Frautschi plots) $\left\langle\frac{1}{r^2}\right\rangle_L$ should fall off a $\frac{1}{2L+1}$ as L increases. This can be achieved (Dalitz) with a harmonic oscillator potential

$$V(r) = V_0 + \lambda r^2 \qquad 9.12$$

This will give a splitting for different L values. For each L value however, there should be four degenerate states which result from combining L with the total quark spin $S = 0$ or $S = 1$. This degeneracy however is removed by spin-orbit splitting

$$\vec{V}_{so} = (\underset{\sim}{\sigma}+\bar{\underset{\sim}{\sigma}})\cdot\underset{\sim}{L}\ V_{so}(r) = \underset{\sim}{S}\cdot\underset{\sim}{L}\ V_{so}(r) \qquad 9.13$$

with mass splitting terms

$$\underset{\sim}{S}\cdot\underset{\sim}{L} = \frac{1}{2}|J(J+1)-L(L+1)-S(S+1)| \qquad 9.14$$

In addition, there is splitting due to spin-spin interaction; this manifests itself mainly in the $L = 0$ states, namely the η-π mass differences. On top of all this, there is SU(3) splitting, giving $q\bar{q} = \{1\} + \{8\}$ for each level. This is illustrated schematically at page 91, for $L = 0,1,2$, although boson clusters for L up to 5 or 6 have been seen. For the baryons, there are two orbital angular momenta to be considered,

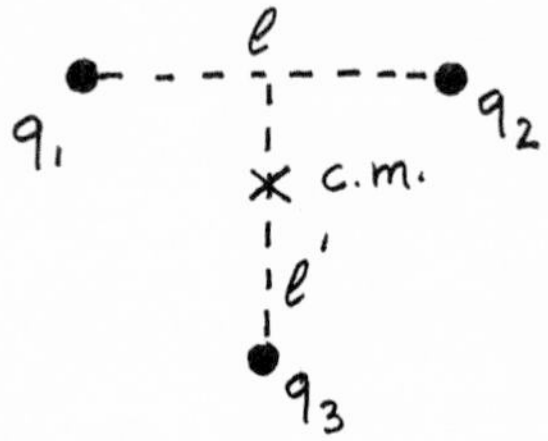

$$\underset{\sim}{L} = \underset{\sim}{\ell} + \underset{\sim}{\ell}' \qquad 9.15$$

The ground state of the baryons, {56} has $L = 0$, parity +. The mass splitting between $\{8_2\}$ and $\{10_4\}$ is just caused by spin-spin splitting between the quarks. For a harmonic oscillator potential, the first excited state has $L = 1$. This yields baryons of negative parity, in a {70} representation of SU(6). For higher excitations, there are of course several combinations of ℓ, ℓ' which can be made.

MESON SPECTROSCOPY

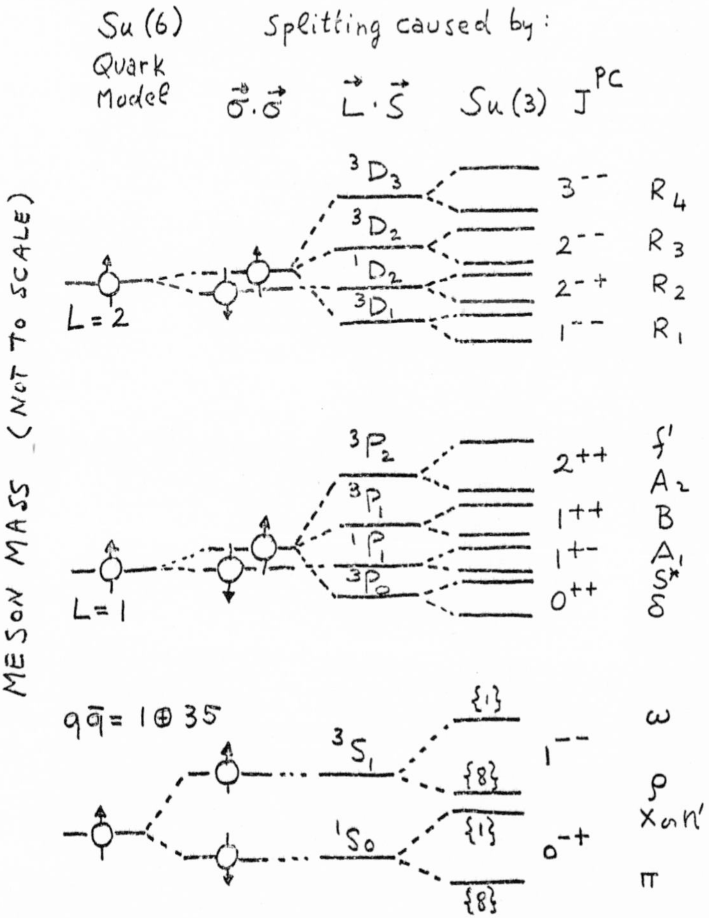

Total L	0^+	1^-	2^+	0^+	$2^+, 1^+, 0^+$
Configuration	$(1s)^2$	(1s)(1p)	(1s)(1d)	(1s)(2s)	$(1p)^2$
Symmetry	s	m	2^+=s,m;	0^+=s,m;	1^+=a
SU(6)	{56}	{70}	{56},{70};	{56},{70};	{20} etc.

This situation, for the lowest states, is illustrated on page . The filled hexagons, triangles, circles, correspond to octets, decuplets, singlets which have been filled with observed states.

This concludes our introduction to particle spectroscopy.

References

F. Gürsey and L.A. Radicati, Phys. Rev. Letters, 13, 173 (1964).

R.H. Dalitz, Symmetries and the Strong Interactions, Proc. XIII Int. Conf. on High Energy Physics at Berkeley. University of California Press, 1967.

B.T. Feld. The Quark model of the Elementary Particles. CERN Report 67-21. Also Models of Elementary Particles. Blaisdell Publ. Co., 1969.

G. Morpurgo. Lectures on the Quark Model, given at the Int. School of Physics, E. Maiorama in Erice, July 1968.

O.W. Greenberg. Resonance Models. Proc. Lund Int. Conf. on Elementary Particles, Institute of Physics, Lund, Sweden, 1969.

V.F. Weisskopf. $SU_2 \rightarrow SU_3 \rightarrow SU_6$. CERN Report 66-19.

BARYON SPECTROSCOPY

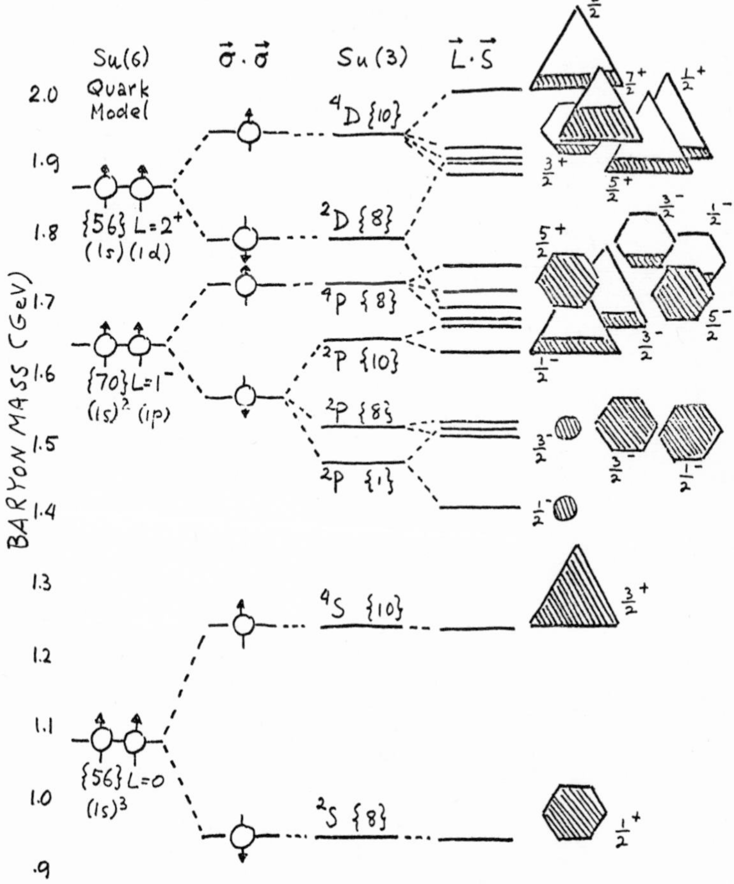

PART I. STRONGLY INTERACTING PARTICLES

B. METHODS IN THE STUDY OF PARTICLES AND RESONANT STATES

INTRODUCTION

There are two main experimental approaches to the study of particles and resonant states:

a) Formation-type experiments:

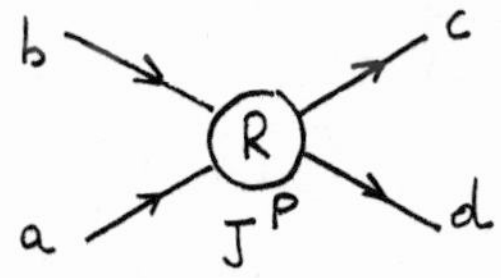

The particle appears as an intermediate state in the reaction

$a + b \to R \to c + d$ (1)

This is also called s-channel or direct channel formation.

It has the following analog in nuclear physics:

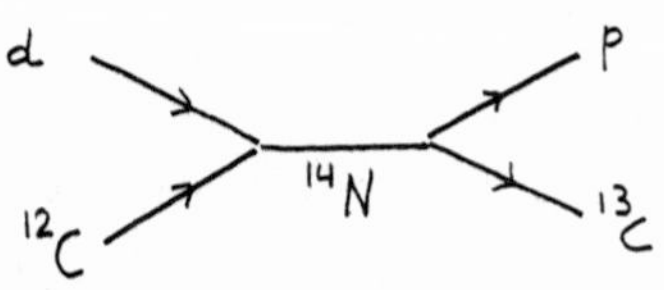

"Compound nucleus" formation

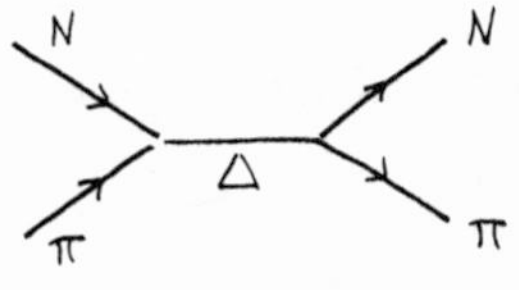

"Direct" or "s-channel" resonance

The total c.m. energy corresponds to the resonant mass.

At $E_{cm} = M_R$, the cross sections for process (1) show enhancements of, e.g., a Breit-Wigner shape. Clearly this approach implies that scattering experiments should be feasible, and thus restricts the type of particles which can be made. Since the target is usually a nucleon, and beams of interest for this purpose π or K mesons, only baryons $N, \Delta, \Lambda, \Sigma, Z(?)$ can be studied.

b) Production-type experiments

This is a two-step process

$$a + b \rightarrow R + c \qquad (2)$$
$$\quad\quad\quad\; \llcorner\; d + e$$

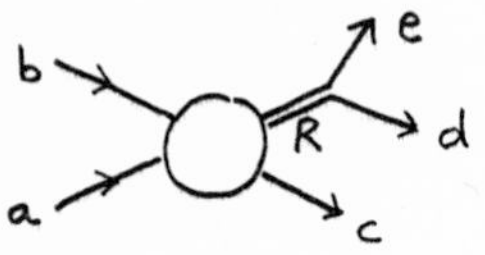

It has also an analog in nuclear physics:

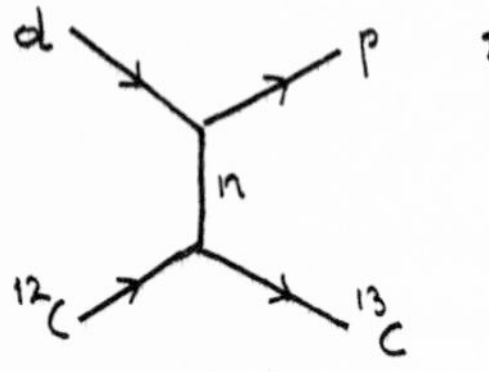

"Direct reaction"

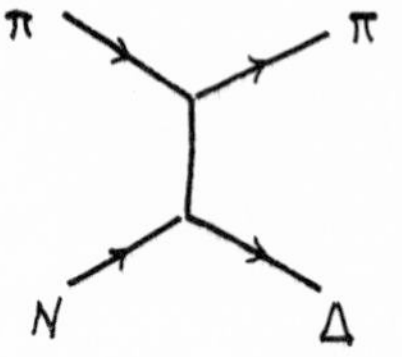

"t-exchange"

The resonance is detected as a peak in the distribution of the invariant mass of two or more particles in the final state.

$$M^2 = -p^{\mu} p^{\mu} = (\sum_i E_i)^2 - (\sum_i \bar{p}_i)^2$$

In this type of process, of course, both mesons and baryons can be produced and studied.

The additive quantum numbers, like hypercharge Y, are determined from the process involved. The isospin I from the observation of the various charge states or from the branching ratios in the decay and reactions, J and P determinations in general require a detailed analysis of the decay angular distributions of the resonant states, together with the polarization of the decay products.

10

ANALYSIS OF FORMATION EXPERIMENTS

Review of Elementary Scattering Theory

a) Partial wave expansion, scattering with absorption. For spinless particles, the scattering of a particle of mass m_1 by a particle of mass m_2, in presence of a potential $V(\bar{r})$ is described by the Schroedinger equation:

$$-\frac{\hbar^2}{2\mu}\nabla^2\psi_f(\vec{r}) + V(\vec{r})\psi_f(\vec{r}) = E\psi_f(\vec{r}) \qquad 10.3$$

where $k = \frac{p}{\hbar} = \frac{\mu v}{\hbar}$; $\mu = \frac{m_1 m_2}{m_1+m_2}$; $\lambda\!\!\!^- = \frac{1}{k}$. For incident beam along the z-axis, the motion is described by the plane wave:

$$\psi_i(r,\theta,\psi) = e^{ikz} \qquad 10.4$$

solution of the equation for a free particle:

$$-\frac{\hbar^2}{2\mu}\nabla^2\psi_i(\vec{r}) = E\psi_i(\vec{r}) \qquad 10.5$$

At larger distance from the scattering center, $r \to \infty$ so the solution of 10.3 is

$$\psi_f(r,\theta) \xrightarrow[r \to \infty]{} e^{ikz} + f(\theta)\frac{e^{ikr}}{r} \qquad 10.6$$

$f(\theta)$ is the scattering amplitude.

Method of partial waves: Expand 10.4, 10.6 in a series of Legendre polynomials:

$$\psi_i = \frac{1}{r} \sum_{\ell=0}^{\infty} v_\ell(r) P_\ell(\cos\theta) \qquad 10.7$$

$$\psi_f = \frac{1}{r} \sum_{\ell=0}^{\infty} u_\ell(r) P_\ell(\cos\theta) \qquad 10.8$$

Where $v_\ell(r)$ and $u(r)$ are the solutions of the radial Schrödinger equations:

$$\frac{d^2}{dr^2} v_\ell(r) + \left[k^2 - \frac{\ell(\ell+1)}{r^2}\right] v_\ell(r) = 0 \quad \text{and} \qquad 9.10$$

$$\frac{d^2}{dr^2} u_\ell(r) + \left[k^2 - U(r) - \frac{\ell(\ell+1)}{r^2}\right] u_\ell(r) = 0 \quad \text{with} \qquad 10.10$$

$U(r) = \frac{2\mu}{\hbar^2} V(\vec{r})$.

If the range R of the potential is finite, the expansion of the incident plane wave, for $r \gg R$, has the following aymptotic behavior:

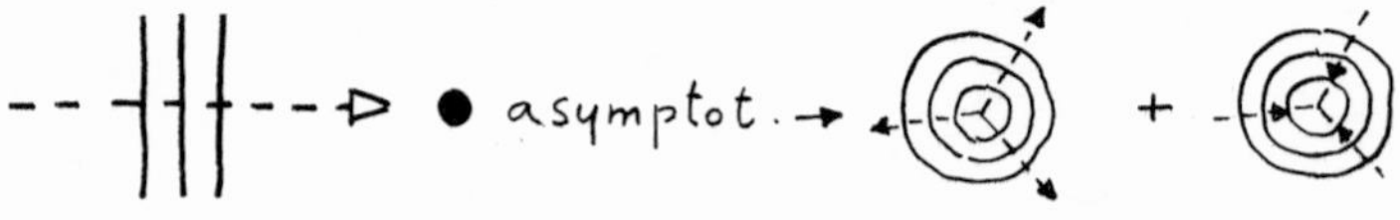

$$\psi_i = e^{ikz} \xrightarrow[kr\to\infty]{} \frac{1}{2ikr} \sum_{\ell=0}^{\infty} i^\ell (2\ell+1) \left[e^{i(kr-\frac{\ell\pi}{2})} - e^{-i(kr-\frac{\ell\pi}{2})} \right] P_\ell(\cos\theta) \qquad 10.11$$

plane wave (↑ e^{ikz}); outgoing (↑ first term); incoming (↑ second term)

spherical wave

Similar expansion of the final w.f., in presence of <u>absorption</u>, yields:

$$\psi_f \to \frac{1}{2ikr}\sum_{\ell=0}^{\infty} i^{\ell}(2\ell+1)\left[\eta_\ell e^{2i\delta_\ell} e^{i(kr-\frac{\ell\pi}{2})} - e^{-i(kr-\frac{\ell\pi}{2})}\right] P_\ell(\cos\theta) \qquad 10.12$$

The effect of the scattering potential, at large distance, is that of altering the outgoing ℓ^{th} wave by a phase shift $2\delta_\ell$ and by an attenuation η_ℓ, if some absorption has taken place.

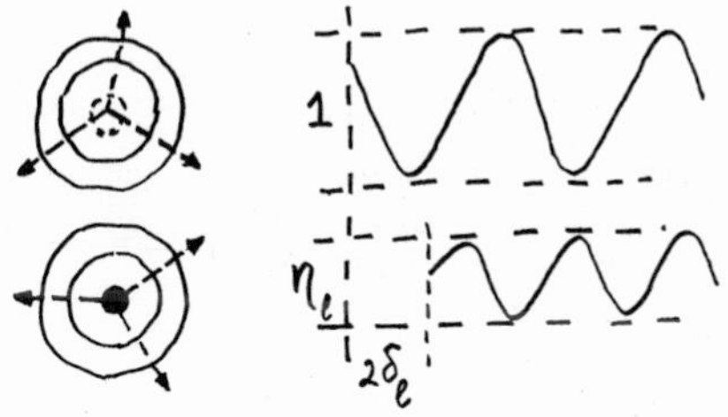

η_ℓ = absorption parameter

δ_ℓ = phase shift

note that $\eta_\ell e^{2i\delta_\ell}$ with η_ℓ, δ_ℓ real is equivalent to $e^{2i\Delta_\ell}$ with Δ_ℓ a complex phase shift.

In fact:

$$e^{2i\Delta_\ell} = e^{2i(\mathrm{Re}\Delta_\ell + i\mathrm{Im}\Delta_\ell)} = e^{-2\mathrm{Im}\Delta_\ell} e^{2i\mathrm{Re}\Delta_\ell} = \eta_\ell e^{2i\delta_\ell} \qquad 10.13$$

where $e^{-2im\Delta_\ell} = \eta_\ell < 1$ for $\mathrm{Im}\Delta_\ell > 0$.

The scattered wave ψ_{scatt} is clearly given by the difference $\psi_f - \psi_i$; subtracting 10.11 from 10.12 we obtain:

$$\psi_{scatt} = \underbrace{\frac{1}{k}\sum_\ell (2\ell+1)\frac{\eta_\ell e^{2i\delta_\ell}-1}{2i} P_\ell(\cos\theta)}_{f(\theta)\ =\ \text{scattering amplitude}} \frac{e^{ikr}}{r} \qquad 10.14$$

We define

$$T_\ell = \frac{\eta_\ell e^{2i\delta_\ell} - 1}{2i} \tag{10.15}$$

as the <u>partial wave</u> amplitude of the ℓth wave. Clearly, for $\eta_\ell = 1$, 10.15 describes purely elastic scattering and can be expressed in the familiar forms:

$$T_\ell\Big)_{\text{one channel}} = \frac{e^{2i\delta_\ell} - 1}{2i} = e^{i\delta_\ell} \sin \delta_\ell = \frac{1}{\cot\delta_\ell - i} \tag{10.16}$$

All the angular dependence of ψ_{scatt} is contained in $f(\theta)$ of 10.14. The <u>differential</u> cross section is then (elastic scattering)

$$\frac{d\sigma_\ell}{d\Omega} = |f(\theta)|^2 = \lambda^2 \left| \sum_\ell (2\ell+1) \left(\frac{\eta_\ell e^{2i\delta_\ell} - 1}{2i} \right) P_\ell(\cos\theta) \right|^2 \tag{10.17}$$

The <u>partial elastic</u> scattering cross section can be obtained either by integrating 10.17

$$\sigma_\ell = 2\pi \int_0^\pi |f(\theta)|^2 d \cos\theta = 4\pi\lambda\!\!\!^-{}^2 \sum (2\ell+1) \left| \frac{\eta_\ell e^{2i\delta_\ell} - 1}{2i} \right|^2 \tag{10.18}$$

or, in equivalent manner, from

$$\sigma_\ell = \frac{N_\ell}{N} = \frac{\text{flux}(\psi_{scatt})}{\text{flux}(\psi_i)} = \frac{\frac{\hbar}{2i\mu} \int_s \left(\frac{\partial \psi_{sc}}{\partial r} \psi^*_{sc} - \frac{\partial \psi^*_{sc}}{\partial r} \psi_{sc} \right) ds}{\text{flux of } e^{ikz} = v_{in} = \frac{\hbar k}{\mu}}$$

$$= \frac{4\pi\lambda\!\!\!^-{}^2 v_{out}}{v_{in}} \sum_\ell (2\ell+1) \left| \frac{\eta_\ell e^{2i\delta_\ell} - 1}{2i} \right|^2 . \quad \text{Here } v_{out} = v_{in}.$$

In conclusion, the partial <u>elastic</u> cross section is

$$\sigma_\ell = 4\pi\lambda\!\!\!^-{}^2 \sum_\ell (2\ell+1) |T_\ell|^2 \tag{10.19}$$

The reaction cross section is determined by the number N_{ab} of particles taken out of the beam per second. N_{ab} equals the incoming flux into a sphere surrounding the scattering center of particles which do not re-emerge through the entrance channel. It will be given by the flux of the complete w.f. ψ_f 10.12:

$$\sigma_r = \frac{N_{ab}}{N} = \frac{\text{flux}(\psi_f)}{\text{flux}(\psi_i)} = \pi\lambda\!\!\!^{-2}\Sigma_\ell(2\ell+1)(1-\eta_\ell^2) \qquad 10.20$$

Consider now the ℓth partial wave only. Its contribution to σ_ℓ and σ_r is:

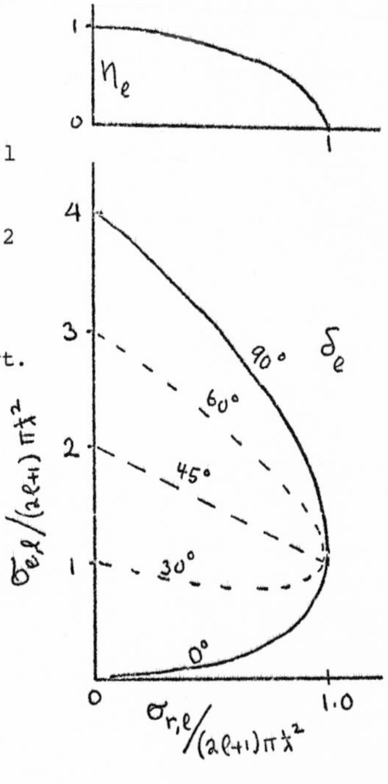

$$\sigma_{e,\ell} = 4\pi\lambda\!\!\!^{-2}(2\ell+1)\left|\frac{\eta_\ell e^{2i\delta_\ell}-1}{2i}\right|^2 \qquad 10.21$$

$$\sigma_{r,\ell} = \pi\lambda\!\!\!^{-2}(2\ell+1)(1-\eta_\ell)^2 \qquad 10.22$$

The relation between $\sigma_{e,\ell}$ and $\sigma_{r,\ell}$ is illustrated in the insert.

For $\eta_\ell = 1$, $\sigma_{r,\ell}$ vanishes and $\sigma_{e,\ell}$ reaches its max. value

$$\sigma_{e,\ell}\Big|_{max} = 4\pi\lambda\!\!\!^{-2}(2\ell+1) \qquad 10.23$$

For $\eta_\ell = 0$, we have maximum absorption (Black disk)

$$\sigma_r = \sigma_{sc} = (2\ell+1)\pi\lambda\!\!\!^{-2} \qquad 10.24$$

Scattering without reaction can occur, but there cannot be reaction without scattering.

Note the factor 4 for pure elastic scattering. This is due to the fact that the incoming and outgoing waves are coherent and can interfere. In this case the interference is constructive. Note also that $(2\ell+1)\pi\lambda\!\!\!^{-2}$ corresponds to the cross sectional area of the ℓth annular region. Note also that $|\eta_\ell|^2 \leq 1$, otherwise the reaction cross section would become negative, i.e., the outgoing wave in 10.12 would have a larger amplitude than the incoming one.

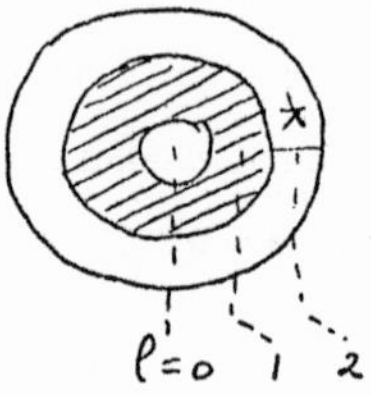

We calculate now the contribution of the ℓth partial wave to the total cross section:

$$\sigma_{T,\ell} = \sigma_{e,\ell} + \sigma_{r,\ell} = 4\pi\lambda\!\!\!^{-2}(2\ell+1)(1+\eta_\ell^2-2\eta_\ell \cos 2\delta_\ell)\frac{1}{4} + \pi\lambda\!\!\!^{-2}(2\ell+1)(1-\eta_\ell^2) \qquad 10.25$$

$$\sigma_{T,\ell} = 2\pi\lambda\!\!\!^{-2}(2\ell+1)(1-\eta_\ell \cos 2\delta_\ell) \qquad 10.26$$

Note now that the imaginary part of $T_\ell = \dfrac{\eta_\ell e^{2i\delta_\ell}-1}{2i}$ is just

$ImT_\ell = \dfrac{1-\eta_\ell \cos 2\delta}{2}$, so that we can write

$$\sigma_{T,\ell} = 4\pi\lambda\!\!\!^{-2}(2\ell+1)(ImT_\ell) \qquad 10.27$$

Adding up the contributions from all partial waves we obtain

$$\sigma_T = 4\pi\bar{\lambda}^2 \sum_\ell (2\ell+1)\,\mathrm{Im}T_\ell \qquad 10.28$$

This is a very important relation. The optical theorem follows at once. Consider $f(\theta)$ of 10.14:

$$f(\theta) = \frac{1}{k}\sum_\ell (2\ell+1)\left(\frac{\eta_\ell e^{2i\delta_\ell}-1}{2i}\right) P_\ell\ (\cos\theta)$$

and take the imaginary part of $f(\theta)$ at $\theta = 0$:

$$\mathrm{Im}\ f(0) = \frac{1}{k}\sum_\ell (2\ell+1)\,\mathrm{Im}T_\ell \qquad 10.29$$

Comparison with 10.27 gives, i.e., the optical theorem.

$$\boxed{\sigma_{tot} = \frac{4\pi}{k}\ \mathrm{Im}\ f(0)} \qquad 10.30$$

Use of the optical theorem enables a determination of the magnitude of the real part of the forward scattering amplitude, from the knowledge of σ_{tot} and $d\sigma/d\Omega(0^\circ)$. In fact,

$$\left.\frac{d\sigma}{d\Omega}\right]_{0^\circ} = |f(0)|^2 = (\mathrm{Re}f(0))^2 + (\mathrm{Im}f(0))^2 \qquad 10.31$$

Since by 10.30

$$(\mathrm{Im}f(0))^2 = \frac{\sigma_T^2}{(4\pi\bar{\lambda})^2} \qquad 10.32$$

$$(\mathrm{Re}f(0))^2 = \left.\frac{d\sigma}{d\Omega}\right]_{0^\circ} - \frac{\sigma_t^2}{(4\pi\bar{\lambda})^2} \qquad 10.33$$

11

SCATTERING FROM PARTICLES WITH SPIN

Density Matrix

We consider here the scattering of mesons, $J^P = 0^-$, on nucleons, $J^P = 1/2^+$. A brief reminder of basic results of the spin density matrix formalism first: Description of a beam of spin 1/2 particles.

$$\psi = \begin{pmatrix} a_1 \\ a_2 \end{pmatrix} = a_1\begin{pmatrix} 1 \\ 0 \end{pmatrix} + a_2\begin{pmatrix} 0 \\ 1 \end{pmatrix}; \quad \psi = \Sigma\, a_i \chi_i \qquad 11.1$$

e.g., beam of free particles of momentum $\hbar k$:

$a_1 = \frac{A_1}{\sqrt{V}}\, e^{i\vec{k}\cdot\vec{r}}$; $a_2 = \frac{A_2}{\sqrt{V}}\, e^{i\vec{k}\cdot\vec{r}}$. Intensity of beam:

$$I = |a_1|^2 + |a_2|^2 \qquad 11.2$$

This can be written as

$$I = (a_1 + a_2)\begin{pmatrix} a_1^* \\ a_2^* \end{pmatrix} = \Sigma_i \psi\psi^\dagger \qquad 11.3$$

Polarization of beam in z-direction:

$$P_z = \frac{|a_1|^2 - |a_2|^2}{|a_1|^2 + |a_2|^2} = \frac{|a_1|^2 - |a_2|^2}{I} \qquad 11.4$$

Define density matrix

$$\rho_{ij} = \psi_i \psi_j^\dagger; \quad \rho = \begin{pmatrix} a_1 a_1^* & a_1 a_2^* \\ a_2 a_1^* & a_2 a_2^* \end{pmatrix} \qquad 11.5$$

$\rho_{ij} = a_i a_j^*$, in fact:

$$\psi\psi^\dagger = (a_i \chi_i)(a_j \chi_j)^\dagger = a_i \chi_i \chi_j^* a_j^* = a_i a_j^* \qquad 11.6$$

The intensity

$$I = |a_1|^2 + |a_2|^2 = \Sigma\, \rho_{ii} = \mathrm{Tr}\, \rho; \qquad \rho = \rho^\dagger \qquad 11.7$$

ρ is Hermitian. Then P_z can be written as

$$P_z = \frac{\mathrm{Tr}\, (\rho\sigma_z)}{\mathrm{Tr}\, \rho} = \frac{\mathrm{Tr}\left[\rho\begin{pmatrix}1 & 0\\ 0 & -1\end{pmatrix}\right]}{\mathrm{Tr}\, \rho} \qquad 11.8$$

Fully polarized beam: $\rho = \begin{pmatrix}I & 0\\ 0 & 0\end{pmatrix}$ or $\rho = \begin{pmatrix}0 & 0\\ 0 & I\end{pmatrix}$

Unpolarized beam: $$\rho = \frac{1}{2}I \begin{pmatrix}1 & 0\\ 0 & 1\end{pmatrix} \qquad 11.9$$

Incoherent superposition of several beams:

$$\begin{cases} \bar\rho_{ij} = \Sigma_\alpha \rho_{ij}^\alpha \omega(\alpha) = \overline{\psi_i \psi_j^\dagger} \\ \text{Intensity} \quad I = \mathrm{Tr}(\bar\rho) = \Sigma_\alpha\, \omega(\alpha) \end{cases} \qquad 11.10$$

This is an average over all states, each with density matrix ρ_{ij}^α, weight $\omega(\alpha)$. In general the expectation value of an operator 0 is given by

$$\langle 0 \rangle = \frac{\mathrm{Tr}\, (0\bar\rho)}{\mathrm{Tr}\, \bar\rho} \qquad 11.11$$

How to construct density matrix: Express $\bar\rho$ as linear combination of Pauli + Unit matrix:

$$\bar\rho = a\begin{pmatrix}1 & 0\\ 0 & 1\end{pmatrix} + b_x\begin{pmatrix}0 & 1\\ 1 & 0\end{pmatrix} + b_y\begin{pmatrix}0 & -i\\ i & 0\end{pmatrix} + b_z\begin{pmatrix}1 & 0\\ 0 & -1\end{pmatrix} \qquad 11.12$$

$$\bar\rho_{11} = a + b_z,\ \rho_{12} = b_x - ib_y;\ \rho_{21} = b_x + ib_y,$$
$$\rho_{22} = a - b_z,\ \bar\rho = a\cdot\mathbb{1} + \vec{b}\cdot\vec{\sigma} \qquad 11.13$$

$$\mathrm{Tr}\rho = a + b_z + a - b_z = I = 2a \qquad 11.14$$

$$\langle\sigma_x\rangle = \frac{\mathrm{Tr}\begin{pmatrix}0 & 1\\ 1 & 0\end{pmatrix}\begin{pmatrix}a+b_z & b_x-ib_y\\ b_x+ib_y & a-b_z\end{pmatrix}}{2a} = \frac{b_x}{a}$$

$$\langle\sigma_y\rangle = \frac{b_y}{a}; \quad \langle\sigma_z\rangle = \frac{b_z}{a}; \qquad 11.15$$

So, the vector $\vec{b}$ is directed along the expectation value of the spin. The polarization vector is $\vec{\rho} = \frac{\vec{b}}{a}$. Then

$$\rho = \frac{1}{2}I(\mathbb{1} + \vec{\rho}\cdot\vec{\sigma}) \qquad 11.16$$

E.g., completely polarized beam in $+z$ direction, like in scattering of mesons on polarized proton target:

$\rho = \frac{1}{2}I(\mathbb{1} + \sigma_z) = \begin{pmatrix}I & 0\\ 0 & 0\end{pmatrix}$ as seen before.

Unpolarized beam (on target)

$\rho = \frac{1}{2}I(\mathbb{1}) = \frac{1}{2}I\begin{pmatrix}1 & 0\\ 0 & 1\end{pmatrix}$ ($\vec{\rho}$ vanishes in any direction)

We want to investigate now the effect of the scattering process on the state of polarization of the incident beam (e.g., $\frac{1}{2}^+ \to 0^+$), or conversely, how the polarization of the target affects the scattering of incident spin zero particles ($0^- \to \frac{1}{2}^+$). In the c.m. these two cases are clearly equivalent. The first corresponds to, e.g., n + He scattering, the second, $\pi, K + p$ scattering. With the inclusion of spin, the asymptotic solution of the Schrödinger equation given in 10.6 takes the form:

$$\psi(r,\theta,\psi) = e^{ikz}\psi^m + \frac{e^{ikr}}{r} M\psi^m \qquad 11.17$$

where M is a scattering amplitude matrix. M transforms

the initial spin state ψ_{in} into a final spin state ψ_f

$$M\psi_{in} = \psi_f; \quad \psi_f^\dagger = \psi_{in}^\dagger M^\dagger \qquad 11.18$$

The final state density matrix can be calculated from the initial density matrix:

$$\rho_{\ell m}^{(i)} = \psi_\ell^{(i)} \psi^{(i)\dagger}; \quad \rho_{\ell m}^{(f)} = \psi_\ell^{(f)} \psi_m^{(i)\dagger};$$
$$\rho_{\ell m}^{(f)} = \psi_f \psi_f^\dagger = M\psi_\ell^{(i)} \psi_m^{(i)} M^\dagger = M\rho_{\ell m}^{(i)} M^\dagger \qquad 11.19$$

so

$$\rho_f = M\rho_i M^\dagger \qquad 11.20$$

The differential scattering cross section, for unit incident intensity, will be

$$\frac{d\sigma}{d\Omega} = \frac{\text{Tr}\ \rho_f}{\text{Tr}\ \rho_i} = \frac{\text{Tr}\ M\rho_i M^+}{\text{Tr}\ \rho_i} \qquad 11.21$$

For example, for an initially unpolarized beam of unit intensity:

$$\rho_i = \rho_0 = \frac{1}{2}\begin{pmatrix} 1 & 0 \\ 0 & 1 \end{pmatrix}; \quad \rho_f = \rho_1 = M\rho_0 M^\dagger = \frac{1}{2}MM^\dagger \qquad 11.22$$

We must specify now the scattering amplitude matrix. M will be in this case a 2 × 2 matrix. Angular momentum and parity conservation require that the matrix elements of M be true scalars, invariant under rotation and reflection. The vectors at disposal are $\vec{\sigma}, \vec{\rho}_i, \vec{\rho}_f$. As it will be shown in detail later, the appropriate form of M is

$$\boxed{M = g(\theta) + ih(\theta)\vec{\sigma}\cdot\hat{n}} \qquad 11.23$$

where $\hat{n}$ is the versor normal to the scattering plane $\hat{n} = \bar{p}_i \times \bar{p}_f / |\bar{p}_i \times \bar{p}_f|$.

- - - - - - - - - -

Note: Often M is written as $M = f(\theta) + g(\theta)\vec{\sigma}\cdot\hat{n}$; our choice is consistent with the results of a treatment of the scattering problem in Born approximation, with an optical model potential and a spin-orbit coupling term. In such a treatment, the term in M which depends on $\vec{\sigma}\cdot\vec{L}$, ultimately on $\vec{\sigma}\cdot\hat{n}$, is purely imaginary.

- - - - - - - - - -

The explicit form of M will, of course, depend on the choice of the axis z of quantization. If we make the choice of z <u>along the direction of the incident beam</u>, the reason for calling $h(\theta)$ the <u>spin-flip</u> amplitude becomes apparent. In this case, in fact,

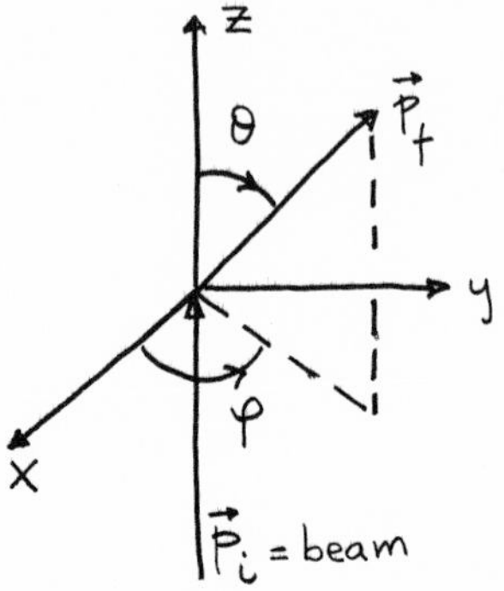

$$\hat{n} = (-\sin\phi, +\cos\phi, 0) \tag{11.24}$$

$$\vec{\sigma}\cdot\hat{n} = -\begin{pmatrix} 0 & 1 \\ 1 & 0 \end{pmatrix}\sin\phi + \begin{pmatrix} 0 & -i \\ i & 0 \end{pmatrix}\cos\phi = \begin{pmatrix} 0 & -ie^{-i\phi} \\ ie^{i\phi} & 0 \end{pmatrix} \tag{11.25}$$

$$M = \begin{pmatrix} g & he^{-i\phi} \\ -he^{i\phi} & g \end{pmatrix} \tag{11.26}$$

For example: $M\Psi_+ = M\begin{pmatrix} 1 \\ 0 \end{pmatrix} = \begin{pmatrix} g \\ -he^{i\psi} \end{pmatrix}$ etc. For scattering taking place in the <u>x-z</u> plane, $\psi = 0$, and

$$M = \begin{pmatrix} g & h \\ -h & g \end{pmatrix}; \quad M\begin{pmatrix} 1 \\ 0 \end{pmatrix} = \begin{pmatrix} g \\ -h \end{pmatrix} = g\begin{pmatrix} 1 \\ 0 \end{pmatrix} - h\begin{pmatrix} 0 \\ 1 \end{pmatrix}$$
$$M\begin{pmatrix} 0 \\ 1 \end{pmatrix} = \begin{pmatrix} h \\ g \end{pmatrix} = h\begin{pmatrix} 1 \\ 0 \end{pmatrix} + g\begin{pmatrix} 0 \\ 1 \end{pmatrix} \tag{11.27}$$

Clearly $h(\theta)$ is always associated with the spin state which has flipped, thus the name <u>spin-flip</u> amplitude. Now we can use the density matrix formulation and derive expressions for the intensity and polarization after the scattering process. If the target of spin $\frac{1}{2}$, or beam, are initially unpolarized, we have from 11.21, 11.22, 11.27:

$$I_f(\theta) = d\sigma/d\Omega = \tfrac{1}{2}\mathrm{Tr}MM^\dagger = |g|^2 + |h|^2 \qquad 11.28$$

The polarization after scattering will be, from 11.11:

$$P_f(\theta) = \langle \sigma_n \rangle = \frac{\mathrm{Tr}(\rho_1\sigma_n)}{\mathrm{Tr}\rho_1} = \frac{\frac{1}{2}\mathrm{Tr}MM^+\sigma_n}{|g|^2+|h|^2} =$$

$$= \frac{\frac{1}{2}\mathrm{Tr}(g+ih\sigma_n)(g^*-ih^*\sigma_n)\sigma_n}{|g|^2 + |h|^2} = \qquad 11.29$$

$$= \frac{\frac{1}{2}\, 2\ \mathrm{Im}(gh^*)\mathrm{Tr}\sigma_n\sigma_n}{|g|^2 + |h|^2} = 2\ \mathrm{Im}(gh^*)/I_f(\theta)$$

In conclusion:

$$I(\theta) = |g|^2 + |h|^2$$
$$I(\theta)\vec{P}(\theta) = 2\ \mathrm{Im}gh^*\hat{n} \qquad 11.30$$

Note that $2\ \mathrm{Im}gh^* = -2\ \mathrm{Im}g^*h$. An important case to be considered is the scattering of mesons from a polarized proton target. With the same choice of z-axis as above (page 110), scattering in the x-z plane, and initial target polarization $\vec{P}_T$ along the y-axis, the differential cross section takes the form:

$$I_f(\pm\theta) = I_f(\theta)(1\pm P_T P_f) \qquad 11.31$$

where $I_f(\theta)$ is the differential cross section for the unpolarized case. $P_f(\theta)$ is, as before, the target polarization after scattering. In general

$$I_f(\pm\theta) = I_f(\theta)(1 \pm \vec{P}_T(\vec{\sigma}\cdot\hat{n}) \qquad 11.32$$

By defining scattering at an angle θ as $I(\theta) = R$, and $I(-\theta) = L$, the "asymmetry parameter" is

$$\frac{L-R}{L+R} = \varepsilon = P_T P_f \qquad 11.33$$

Thus the final proton polarization can be obtained from the measurement of the left-right asymmetry and the knowledge of the initial target polarization. Proof of the above relations will be a homework assignment.

Partial Wave Analysis of Meson-Nucleon Scattering

Consider the scattering of π or K mesons (0^-) on protons. For orbital angular momentum ℓ, scattering can occur in the j-states:

$$j_+ = \ell + \frac{1}{2} \text{ or } j_- = \ell - \frac{1}{2} \qquad 11.34$$

We make use of the projection operators P_+, P_- defined as

$$P_+|j_+\rangle = |j_+\rangle; \quad P_-|j_-\rangle = |j_-\rangle; \quad P_+^2 = P_-^2 = 1; \qquad 11.35$$

$$P_+|j_-\rangle = 0; \quad P_-|j_+\rangle = 0;$$

we also have

$$\vec{J} = \vec{L} + \frac{\vec{\sigma}}{2}; \quad \langle J^2 \rangle = j(j+1) = \ell(\ell+1) + \frac{3}{4} + \langle \vec{L}\cdot\vec{\sigma} \rangle \qquad 11.36$$

$$\langle \vec{L}\cdot\vec{\sigma} \rangle = j(j+1) - \ell(\ell+1) - \frac{3}{4} = \begin{cases} \ell & \text{for } j_+ = \ell+\frac{1}{2} \\ -\ell-1 & \text{for } j_- = \ell-\frac{1}{2} \end{cases} \qquad 11.37$$

Then

$$P_+ = \frac{\ell+1+\vec{L}\cdot\vec{\sigma}}{2\ell+1} \text{ has eigenvalues } \begin{matrix} +1 & \text{for } j_+ \\ 0 & \text{for } j_- \end{matrix}$$

$$P_- = \frac{\ell-\vec{L}\cdot\vec{\sigma}}{2\ell+1} \quad " \quad " \quad \begin{matrix} +1 & \text{for } j_- \\ 0 & \text{for } j_+ \end{matrix} \qquad 11.38$$

Express $\vec{\sigma}\cdot\vec{L}$ as follows:

$$\vec{L} = i\vec{r} \times \vec{\nabla} = -i\vec{r} \times \hat{\ell}_\theta \frac{1}{r} \sin\theta \frac{d}{d(\cos\theta)} \qquad 11.39$$

($\hat{e}_\theta$ is a unit vector along increasing θ; since the process is parity conserving, there is no ϕ dependence.) Now since $\vec{r} \times \hat{e} \frac{1}{r} = -\hat{n}$, we have

$$\vec{\sigma}\cdot\vec{L} = i\vec{\sigma}\cdot\hat{n} \sin\theta \frac{d}{d\cos\theta} \qquad 11.40$$

We write now the asymptotic expansions of 11.17:

$$e^{ikz}\Psi^m \underset{r\to\infty}{\longrightarrow} \frac{1}{2ikr}\sum_{\ell=0}^{\infty} i^\ell(2\ell+1)\left[e^{i(kr-\frac{\ell\pi}{2})} - e^{-i(kr-\frac{\ell\pi}{2})}\right](P_+ + P_-)P_\ell(\cos\theta)\Psi^m \qquad 11.41$$

$$\Psi_f \underset{r\to\infty}{\longrightarrow} \frac{1}{2ikr}\sum_{\ell=0}^{\infty} i^\ell(2\ell+1)\left[\eta_\ell e^{2i\delta_\ell} e^{i(ir-\frac{\ell\pi}{2})} - e^{-i(kr-\frac{\ell\pi}{2})}\right](P_+ + P_-)P_\ell(\cos\theta)\Psi^m \qquad 11.42$$

$\Psi_{scatt} = \Psi_f - \Psi_i =$

$$\frac{e^{ikr}}{r}\frac{1}{k}\sum_{\ell=0}^{\infty}(2\ell+1)\left[\frac{\eta_{\ell_+} e^{2i\delta_{\ell_+}}-1}{2i} P_+ + \frac{\eta_{\ell_-} e^{2i\delta_{\ell_-}}-1}{2i} P_-\right]P_\ell(\cos\theta)\Psi^m \qquad 11.43$$

$$\Psi_{scatt} = \frac{e^{ikr}}{r} \frac{1}{k}\sum_{\ell=0}^{\infty}(2\ell+1)\left[T_{\ell_+}P_+ + T_{\ell_-}P_-\right]P_\ell(\cos\theta)\Psi^m \qquad 11.44$$

where

$$T_{\ell_\pm} = \frac{\eta_{\ell_\pm} e^{2i\delta_{\ell_\pm}}-1}{2i} \qquad 11.45$$

is the partial wave amplitude for scattering in states $\ell \pm \frac{1}{2}$.

Making use now of P_+ and P_-, 11.38,

$$\Psi_{scatt} = \frac{e^{ikr}}{r} \cdot \frac{1}{k} \sum_{\ell=0}^{\infty} \left\{ \left[(\ell+1) T_{\ell_+} + \ell T_{\ell_-} \right] P_\ell(\cos\theta) + \left[i\vec{\sigma}\cdot\hat{n} \left(T_{\ell_+} - T_{\ell_-} \right) \frac{d}{d\cos\theta} P_\ell(\cos\theta) \right] \right\} \psi^m \qquad 11.46$$

We can see that 11.46 gives for the scattering amplitude matrix M, the form conjectured previously:

$M = g(\theta) + ih(\theta)\vec{\sigma}\cdot\hat{n}$, where

$$g(\theta) = \frac{1}{k} \Sigma_\ell \left[(\ell+1) T_{\ell_+} + \ell T_{\ell_-} \right] P_\ell(\cos\theta)$$

$$h(\theta) = \frac{1}{k} \Sigma_\ell \left(T_{\ell_+} - T_{\ell_-} \right) \sin\theta \frac{d}{d\cos\theta} P_\ell(\cos\theta) \qquad 11.47$$

The effect of the spin-orbit interaction is entirely contained in the spin-flip amplitude $h(\theta)$, when different from zero and if $T_{\ell_+} \neq T_{\ell_-}$. For $T_{\ell_+} = T_\ell \rightarrow h(\theta) = 0$,

$g(\theta) \equiv f(\theta) = \frac{1}{k} \Sigma_\ell (2\ell+1) T_\ell P_\ell(\cos\theta)$ as in 10.14, valid for the scattering of spinless particles.

The differential angular distribution or the polarization can now be explicitly expanded into a series of either Legendre polynomials, or powers of $\cos\theta$: with $\cos\theta = \mu$, $T_{\ell\pm} \equiv \ell_{2J}$

$$\left.\begin{aligned} g(\theta) &= \lambda\!\!\!^{-}\{s_1 + (p_1+2p_3)P_1(\mu) + (2d_3+3d_5)P_2(\mu) + \dots \\ h(\theta) &= \lambda\!\!\!^{-}\{(p_3-p_1)(P_1^1(\mu) + (d_5-d_3)P_2^1(\mu) + \dots \end{aligned}\right\} \qquad 11.48$$

$$\frac{d\sigma}{d\Omega} = I_f(\theta) = |g|^2 + |h|^2 = \lambda\!\!\!^{-2} \Sigma_n A_n P_n(\mu)$$

$$I_f(\theta) P_f(\theta) = 2 \mathrm{Im} gh^* = \lambda\!\!\!^{-2} \Sigma_n B_n P_n'(\mu) \qquad 11.49$$

A tabulation of A_n and B_n is enclosed. From the expansion of 11.49, note that

$$\sigma = \int \frac{d\sigma}{d\Omega}\, d\Omega = 4\pi \lambda\!\!\!^{-2} A_0 ; \quad A_0 = \frac{\sigma}{4\pi \lambda\!\!\!^{-2}} \qquad 11.50$$

Additional comments. The explicitly covariant helicity formalism of Jacob and Wick yields 11.47 for $0^- \frac{1}{2}^+ \rightarrow 0^- \frac{1}{2}^+$ in the CM frame. If the interaction occurs in a single eigenstate of J^P, the A_n coefficients (see table) are $\neq 0$ only for order $n \leq 2J - 1$ and n even, and all $B_n = 0$. In this case the angular distribution is symmetric about $\cos \theta = 0$ and there is no polarization.

A determination of partial waves based on $d\sigma/d\Omega$ is affected by the "Minami ambiguity." An interchange of parity for all states, for J constant, corresponds to the so-called Minami transformation:

$$T_\ell^+ \rightleftarrows T_{\ell+1}^- \qquad \text{e.g.,} \quad S_1 \leftrightarrow P_1, \quad P_3 \leftrightarrow D_3, \quad \text{etc.}$$

Formally, this transformation is induced by the pseudoscalar operation $\vec{\sigma}\cdot\vec{k}$, acting on the transition matrix M:

$$M' = \vec{\sigma}\cdot\vec{k}_f \, M \, \vec{\sigma}\cdot\vec{k}_i \qquad 11.51$$

M' is still a scalar, but with respect to M, the initial and final states are transformed into new states of the same J, but opposite parity. If $I_f(\theta)$ and $P_f(\theta)$ are now evaluated,

$$\begin{aligned} I_f'(\theta) &= I(\theta) \\ I_f'(\theta)P_f'(\theta) &= -I_f(\theta)P_f(\theta) \end{aligned} \qquad 11.52$$

Thus the Minami transformation leaves $d\sigma/d\Omega$ invariant but changes the sign of the polarization. In practice the

$$I(\theta) = \lambda\!\!\!^{-2} \sum_n A_n P_n(\cos\theta)$$ [a]

	A_0	A_1	A_2	A_3	A_4	A_5	A_6	A_7	A_8	A_9
$S_1S_1+P_1P_1$	1	--	--	--	--	--	--	--	--	--
S_1P_1	--	2	--	--	--	--	--	--	--	--
$S_1P_3+P_1D_3$	--	4	--	--	--	--	--	--	--	--
$S_1D_3+P_1P_3$	0	--	4	--	--	--	--	--	--	--
$S_1D_3+P_1F_5$	0	--	6	--	--	--	--	--	--	--
$S_1F_5+P_1D_5$	--	0	--	6	--	--	--	--	--	--
$S_1F_7+P_1G_7$	--	0	--	8	--	--	--	--	--	--
$S_1G_7+P_1F_7$	0	--	0	--	8	--	--	--	--	--
$S_1G_9+P_1H_9$	0	--	0	--	10	--	--	--	--	--
$S_1H_9+P_1G_9$	--	0	--	0	--	10	--	--	--	--
$P_3P_3+D_3D_3$	2	--	2	--	--	--	--	--	--	--
P_3D_3	--	4/5	--	36/5	--	--	--	--	--	--
$P_3D_5+D_3F_5$	--	36/5	--	24/5	--	--	--	--	--	--
$P_3F_5+D_3D_5$	0	--	12/7	--	72/7	--	--	--	--	--
$P_3F_7+D_3G_7$	0	--	72/7	--	40/7	--	--	--	--	--
$P_3G_7+D_3F_7$	--	0	--	8/3	--	40/3	--	--	--	--
$P_3G_9+D_3H_9$	--	0	--	40/3	--	20/3	--	--	--	--
$P_3H_9+D_3G_9$	0	--	0	--	40/11	--	180/11	--	--	--
$D_5D_5+F_5F_5$	3	--	24/7	--	18/7	--	--	--	--	--
D_5F_5	--	18/35	--	16/5	--	100/7	--	--	--	--
$D_5F_7+F_5G_7$	--	72/7	--	8	--	40/7	--	--	--	--
$D_5G_7+F_5F_7$	0	--	8/7	--	360/77	--	200/11	--	--	--
$D_5G_9+F_5H_9$	0	--	100/7	--	720/77	--	70/11	--	--	--
$D_5H_9+F_5G_9$	--	0	--	20/11	--	80/13	--	3150/143	--	--
$F_7F_7+G_7G_7$	4	--	100/21	--	324/77	--	100/33	--	--	--
F_7G_7	--	8/21	--	24/11	--	600/91	--	9800/429	--	--
$F_7G_9+G_7H_9$	--	40/3	--	120/11	--	120/13	--	2800/429	--	--
$F_7H_9+G_7G_9$	0	--	200/231	--	3240/1001	--	280/33	--	3920/143	--
$G_9G_9+H_9H_9$	5	--	200/33	--	810/143	--	160/33	--	490/143	--
G_9H_9	--	10/33	--	240/143	--	60/13	--	78400/8203	--	79380/2431

a) Here $S_1P_3+P_1D_3$ stands for $\mathrm{Re}(S_1^*P_3+P_1^*D_3)$, etc.

$$\vec{P}\cdot\hat{n}I(\theta) = \lambda\!\!\!^{-2}\Sigma_n B_n P_n^1(\cos\theta)^a$$

	B_1	B_2	B_3	B_4	B_5	B_6	B_7	B_8	B_9
S_1P_1	2	--	--	--	--	--	--	--	--
$S_1P_3-P_1D_3$	-2	--	--	--	--	--	--	--	--
$S_1D_3-P_1D_3$	--	2	--	--	--	--	--	--	--
$S_1D_5-P_1F_5$	--	-2	--	--	--	--	--	--	--
$S_1F_5-P_1D_5$	0	--	2	--	--	--	--	--	--
$S_1F_7-P_1G_7$	0	--	-2	--	--	--	--	--	--
$S_1G_7-P_1F_7$	--	0	--	2	--	--	--	--	--
$S_1G_9-P_1H_9$	--	0	--	-2	--	--	--	--	--
$S_1H_9-P_1G_9$	0	--	0	--	2	--	--	--	--
P_3D_3	8/5	--	12/5	--	--	--	--	--	--
$P_3D_5-D_3F_5$	-18/5	--	-2/5	--	--	--	--	--	--
$P_3F_5-D_3D_5$	--	10/7	--	18/7	--	--	--	--	--
$P_3F_7-D_3G_7$	--	-24/7	--	-4/7	--	--	--	--	--
$P_3G_7-D_3F_7$	0	--	4/3	--	8/3	--	--	--	--
$P_3G_9-D_3H_9$	0	--	-10/3	--	-2/3	--	--	--	--
$P_3H_9-D_3G_9$	--	0	--	14/11	--	30/11	--	--	--
D_5F_5	54/35	--	8/5	--	20/7	--	--	--	--
$D_5F_7-F_5G_7$	-36/7	--	-2/3	--	-4/21	--	--	--	--
$D_5G_7-F_5F_7$	--	4/3	--	18/11	--	100/33	--	--	--
$D_5G_9-F_5H_9$	--	-100/21	--	-72/77	--	-10/33	--	--	--
$D_5H_9-F_5G_9$	0	--	40/33	--	64/39	--	450/143	--	--
F_7G_7	32/21	--	16/11	--	160/91	--	1400/429	--	--
$F_7G_9-G_7H_9$	-20/3	--	-10/11	--	-12/39	--	-50/429	--	--
$F_7H_9-G_7G_9$	--	100/77	--	1458/1001	--	20/11	--	490/143	--
G_9H_9	50/33	--	200/143	--	20/13	--	14000/7293	--	8820/2431

[a] Here $S_1P_3-P_1D_3$ is to be taken as $\mathrm{Im}(S_1^*P_3-P_1^*D_3)$, etc.; $\hat{n}$ is the production normal, $\vec{P}_i\times\vec{P}_f/|\vec{P}_i\times\vec{P}_f|$ and $P_n^{\,1}(\cos\theta) = \sin\theta(d/d\cos\theta)P_n(\cos\theta)$.

knowledge of the polarization removes this ambiguity. A further ambiguity remains, referred to as the "complex conjugation" ambiguity. This arises due to the property

$$\begin{array}{ll} T \to T^* & I'_f(\theta) = I_f(\theta) \\ \delta_\ell \to -\delta_\ell & P'_f(\theta) = -P_f(\theta) \end{array} \qquad 11.53$$

As can be seen, a Minami transformation followed by complex conjugation leaves $d\sigma/d\Omega$ and $P(\theta)$ invariant.

Dynamical arguments often remove also the second kind of ambiguity, e.g., Coulomb-nuclear interference, Wigner condition near a resonance, or any other argument or measurement giving the sign of the real part of the scattering amplitude.

12

LOW ENERGY BEHAVIOUR OF PARTIAL WAVES

Energy Dependence of Phase Shifts

Will limit here the discussion to those properties which are independent of the particular form of the potential, but which follow from the structure of the Schrödinger equation. At low energies, the behaviour of the phase shifts is determined by the effect of the centrifugal barrier. This can be seen as follows: Consider again the radial Schrödinger equation on page 100. We have so far made use of the asymptotic solution, for $kr \to \infty$, of the form

$$u_\ell \xrightarrow[kr \to \infty]{} c \sin\left(kr - \frac{\ell\pi}{2} + \delta_\ell\right) \qquad \text{or} \qquad 12.1$$

$$u_\ell \xrightarrow[kr \to \infty]{} \cos\delta_\ell \sin\left(kr - \frac{\ell\pi}{2}\right) + \sin\delta_\ell \cos\left(kr - \frac{\ell\pi}{2}\right) \qquad 12.2$$

Assuming a short range potential $V(r)$ of radius r_o, for $kr > kr_o$ the general solution would be of the form:

($V(r) = 0$ in this region)

$$u_\ell \sim \cos\delta_\ell\, J_{\ell+\frac{1}{2}}(kr) + \sin\delta_\ell\, N_{\ell+\frac{1}{2}}(kr) \qquad \text{where} \qquad 12.3$$

$J_{\ell+\frac{1}{2}}(kr)$ = Spherical Bessel function of order $p = \ell + \frac{1}{2}$ (regular solution)

$N_{\ell+\frac{1}{2}}(kr)$ = Neumann function (irregular solution).

The limiting forms of the two functions are:

$$J_{\ell+\frac{1}{2}}(kr) \to \sin\left(kr - \frac{\ell\pi}{2}\right) \quad \text{as} \quad kr \to \infty$$

$$\frac{(kr)^{\ell+1}}{(2\ell+1)!!} \quad \text{as} \quad kr \to 0 \qquad 12.4$$

$$N_{\ell+\frac{1}{2}}(kr) \to \cos\left(kr - \frac{\ell\pi}{2}\right) \quad \text{as} \quad kr \to \infty$$

$$\frac{(2\ell-1)!!}{(kr)^{\ell}} \quad \text{as} \quad kr \to 0$$

We have obviously made use so far of the limiting forms as $kr \to \infty$. We are interested here in the other case, when $kr \to 0$:

$$u_\ell \underset{kr \to 0}{\longrightarrow} \cos\delta_\ell \frac{(kr)^{\ell+1}}{(2\ell+1)!!} + \sin\delta_\ell \frac{(2\ell-1)!!}{(kr)^{\ell}} \qquad 12.5$$

The phase shift δ_ℓ will be determined, aside from the form of the potential, by imposing continuity of the wave function and its derivative at the boundary $R = r_o$ of the interaction region, i.e., by joining the logarithmic derivative of the w.f. at such boundary:

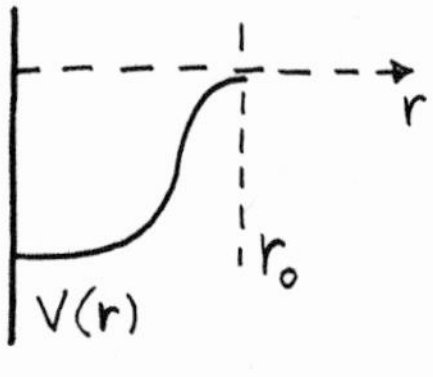

$$f_\ell(k) = r_o\left(\frac{u'}{u}\right)_{r=r_o} \qquad 12.6$$

or

$$f_\ell(k) = \frac{\frac{\ell+1}{(2\ell+1)!!}(kr_o)^{\ell+1}\cos\delta_\ell - \ell\frac{(2\ell-1)!!}{(kr_o)^{\ell}}\sin\delta_\ell}{\frac{(kr_o)^{\ell+1}}{(2\ell+1)!!}\cos\delta_\ell + \frac{(2\ell-1)!!}{(kr_o)^{\ell}}\sin\delta_\ell}$$

$$f_\ell(k) = \frac{\ell + 1 - \ell \tan\delta_\ell \dfrac{(2\ell-1)!!(2\ell+1)!!}{(kr_o)^{2\ell+1}}}{1 + \tan\delta_\ell \dfrac{(2\ell-1)!!(2\ell+1)!!}{(kr_o)^{2\ell+1}}}$$

$$= \frac{\ell + 1 - \ell F \tan\delta_\ell}{1 + F \tan\delta_\ell} \qquad 12.7$$

with $F = \dfrac{(2\ell-1)!!(2\ell+1)!!}{(kr_o)^{2\ell+1}}$ 12.8

Finally

$$\tan\delta_\ell = \frac{(\ell+1)-f_\ell(k)}{\ell + f_\ell(k)}\frac{1}{F} = \frac{\ell + 1 - f_\ell(k)}{\ell + f_\ell(k)}\,\frac{(kr_o)^{2\ell+1}}{(2\ell-1)!!(2\ell+1)!!} \qquad 12.9$$

The continuity of the logarithmic derivative at the boundary now requires $f_\ell(k)$ to be determined by the behaviour of the w.f. for $r < r_o$. Inside the well, at low energy, the w.f. is practically independent of energy. Therefore $f_\ell(k)$ can be taken as constant. Conclusion:

$$\tan\delta_\ell = \alpha_\ell \frac{(kr_o)^{2\ell+1}}{(2\ell-1)!!(2\ell+1)!!} \qquad 12.10$$

with $\alpha_\ell \approx$ constant. Note that

$$\alpha_\ell = \frac{(\ell+1) - f_\ell}{\ell + f_\ell} \qquad 12.11$$

has poles for $f_\ell = -\ell$. In such a case $\tan\delta_\ell \to \infty$ and $\delta_\ell \to \frac{\pi}{2}$. This corresponds to <u>resonant scattering</u>. Will come back to this later. For the moment we note that at low energy only the smallest ℓ-values contribute appreciably to δ_ℓ:

ℓ	$(2\ell+1)!!(2\ell-1)!!$
0	1
2	45
3	1575
---	----

For $kr_o << 1$, $|\delta_\ell| << 1$

$\delta_\ell \approx (kr_o)^{2\ell+1}$ and

$\delta_\ell \to 0$ for $k \to 0$

As a practical rule, the phase shifts δ_ℓ for $\ell > kr_o$ are usually small enough to be ignored.

Scattering Length and Effective Range

The interaction of, e.g., K^- mesons with nucleons, leading to the formation of Λ and Σ states, is described at low energy much in the same way as low energy n-p scattering, making use of the effective range expansion. There are, however, important differences which we wish to point out, due to the presence of strong absorption. We start from the well-known relation

$$k^{2\ell+1} \cot \Delta_\ell = \frac{1}{A_\ell} + \frac{1}{2} r_\ell k^2 + \ldots \tag{12.12}$$

In the presence of absorption, Δ_ℓ is complex, and so will in general be the scattering length A_ℓ and the effective range r_ℓ. (Note that the (-) sign which usually appears in the definition of the scattering length, is incorporated in this definition into A_ℓ). At sufficiently low energy (e.g., K^- mesons for $p_K \lesssim 250$ MeV/c, the scattering can be satisfactorily described by the zero-effective range approximation, which consists in neglecting the second term in the expansion 12.12. In this approximation, for $\ell = 0$,

$$k \cot \Delta = \frac{1}{A} \tag{12.13}$$

where

$$\Delta = \delta + i\gamma$$

$$A = a + ib \tag{12.14}$$

(Remember: $e^{2i\Delta} = \eta e^{2i\delta} = e^{-2\gamma}e^{2i\delta}$).

The elastic partial wave amplitude, in accord with our previous definition, is now:

$$T_\ell = \frac{e^{2i\Delta}-1}{2i} = \frac{e^{i\Delta}(e^{i\Delta}-e^{-i\Delta})}{2i} = \frac{\sin\Delta}{e^{-i\Delta}} = \frac{\sin\Delta}{\cos\Delta - i\sin\Delta}$$

$$= \frac{1}{\cot\Delta - i} \qquad 12.15$$

In terms of 12.13, this becomes

$$T_\ell = \frac{kA}{1 - ikA} = \frac{ka + ikb}{1 + kb - ika} \qquad 12.16$$

Squaring:

$$|T_\ell|^2 = \frac{k^2(a^2+b^2)}{1 + 2kb + k^2(a^2+b^2)}. \qquad 12.17$$

so that the total elastic cross section is

$$\sigma_\ell = 4\pi\lambda\!\!\!^{-2}|T_\ell|^2 = 4\pi \frac{a^2 + b^2}{1 + 2kb + k^2(a^2+b^2)} \qquad 12.18$$

Note that for $k \to 0$, $\sigma_\ell \to 4\pi|A|^2$, which is a well known result for, e.g., s-wave n-p scattering. For the reaction cross section we have (10.22, p. 103)

$$\sigma_r = \pi\lambda\!\!\!^{-2}(1-\eta^2) \qquad 12.19$$

where

$$\eta = |e^{2i\Delta}| = \left|\frac{1 + i\tan\Delta}{1 - i\tan\Delta}\right| = \left|\frac{1 + ikA}{1 - ikA}\right| \qquad 12.20$$

$$\sigma_r = 4\pi\lambda\!\!\!^{-} \frac{b}{1 + 2kb + (a^2+b^2)k^2}. \qquad 12.21$$

This exhibits the 1/v law of low energy nuclear reactions, in fact σ_r goes as $\frac{1}{k}$ for $k \to 0$.

For $b = 0$, $\sigma_r = 0$. Thus the imaginary part of the scattering length describes the absorption process (of course, since it corresponds, because of 12.14, to the imaginary part of Δ).

<u>Argand diagram of a complex scattering length amplitude</u>. Consider again the elastic amplitude 12.16.

$$T_\ell = \frac{ka + ikb}{1 + kb - ika} .$$

For <u>strong absorption</u> channels,

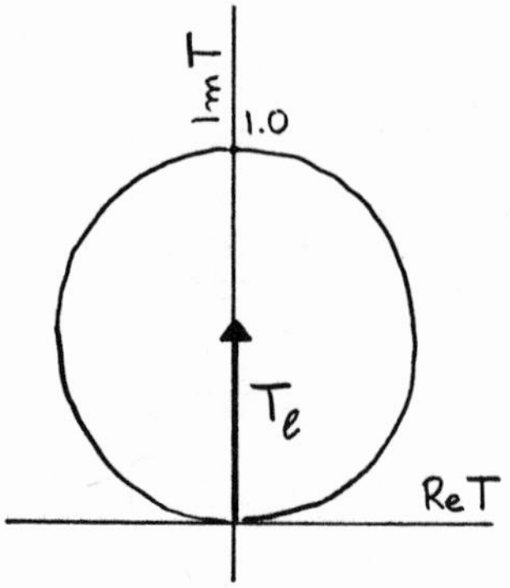

such as in $K^- p \to \Sigma\pi, \Lambda\pi$, $b >> a$ and $T_\ell \approx i \frac{kb}{1 + kb}$ 12.22.

T_ℓ describes a straight line on the imaginary axis, with

$|T_\ell| < \frac{1}{2}$ for $kb < 1$

$|T_\ell| > \frac{1}{2}$ for $kb > 1$.

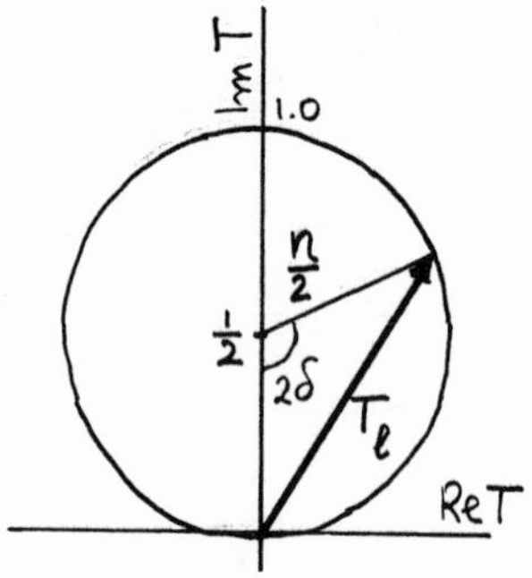

If there is no absorption, $b = 0$, $\eta = 1$ and T_ℓ follows along the unitary circle

$$T_\ell = \frac{ka}{1 - ika} \qquad 12.22.$$

For arbitary A, define

$\alpha = k|A| = k\sqrt{a^2+b^2}$

$\beta = b/a$ 12.23.

From 12.16, lines of β = constant corresponds to circles of radius

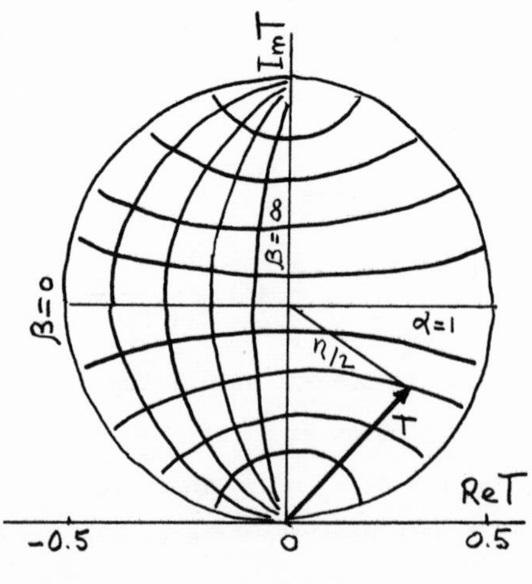

$r = \frac{\sqrt{1+\beta^2}}{2}$ with centres at Re $T_\ell = -\beta/2$, Im $T_\ell = \frac{1}{2}$.

Lines of α = constant correspond to circles of radius

$r = |\alpha/(\alpha^2-1)|$ and centres at Re $T_\ell = 0$, Im $T_\ell = \frac{\alpha^2}{\alpha^2-1}$.

13

COUPLED CHANNELS

S and T Matrices

We reformulate now the partial wave analysis with absorption, to obtain the amplitude for transitions from a given initial channel to a different final channel. We consider, for example,

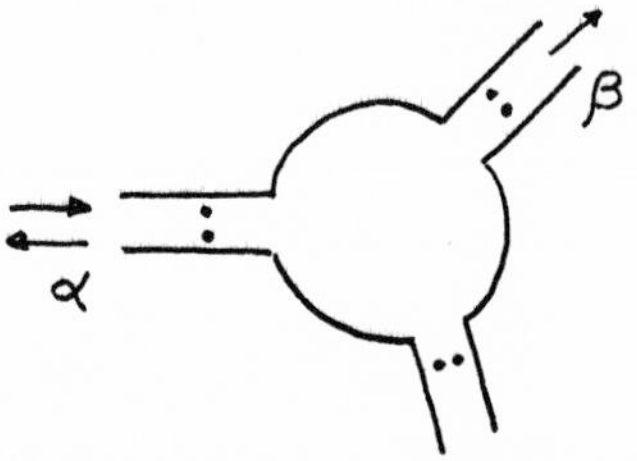

two spin zero particles in each open channel. Since for transitions from one channel to another, the wave number k may be different, one should pay attention here to proper normalization, to ensure conservation of flux:

Normalize so that

Incoming spherical wave: outgoing spherical wave:

$$\frac{1}{\sqrt{v}}\frac{e^{-ikr}}{r} \qquad\qquad \frac{1}{\sqrt{v}}\frac{e^{ikr}}{r}$$

Flux of both is one particle per steradian. We can rewrite the asymptotic wave function describing the relative motion of two particles in a given channel α_1 for values ℓ_α of the orbital angular momentum, as follows:

$$\Psi_{\ell_\alpha}^{(\alpha\to\alpha)}(r) \underset{kr\to\infty}{\sim} \frac{1}{2ik_\alpha r}\left[S_{\alpha\alpha}e^{i(k_\alpha r-\frac{\ell_\alpha\pi}{2})} - e^{-i(k_\alpha r-\frac{\ell_\alpha\pi}{2})}\right] \quad 13.1$$

$S_{\alpha\alpha}$ replaces here $\eta_\ell e^{2i\delta_\ell}$ of our preceding notation. Note incoming + outgoing waves in the same channel α. If there are other channels <u>open</u> (available energy above threshhold) however, with an incoming wave in channel $\underline{\alpha}$, there will be in general outgoing waves in the open channels.

An outgoing wave in channel β is written as

$$\Psi_{\ell_\beta}^{(\alpha\to\beta)}(r) \underset{kr\to\infty}{\sim} \frac{1}{2i\sqrt{k_\alpha k_\beta}\; r} S_{\beta\alpha}e^{i(k_\beta r-\frac{\ell_\beta\pi}{2})} \quad 13.2$$

The cross section for the reaction $\alpha \to \beta$ is given by the ratio of the flux of the outgoing wave in channel β to the incident plane wave flux. For $\ell_\alpha = \ell_\beta$

$$\sigma_{\ell_\alpha}^{(\alpha\to\beta)} = (2\ell_\alpha+1)\frac{\pi}{k_\alpha^2}\left|S_{\beta\alpha}\right|^2, \qquad \text{for } \beta \neq \alpha \quad 13.3$$

The elastic cross section is obtained as done previously, with the result

$$\sigma_\ell^{(\alpha\to\alpha)} = (2\ell_\alpha+1)\frac{\pi}{k_\alpha^2}\left|S_{\alpha\alpha}-1\right|^2 \quad 13.4$$

13.3 and 13.4 can be condensed into

$$\sigma_\ell^{(\alpha\to\beta)} = (2\ell_\alpha+1)\frac{\pi}{k_\alpha^2}\left|S_{\beta\alpha} - \delta_{\beta\alpha}\right|^2 \quad 13.5$$

$S_{\beta\alpha}$, with α,β from 1 to n = number of channels is what is usually called <u>collision</u> or <u>scattering matrix</u>, of order $n \times n$.

Properties of scattering matrix:

1) Unitarity: from 13.2 and 13.3 one can see that for an incoming spherical wave flux of 1 particle/sec in channel the outgoing flux in channel β is $|S_{\beta\alpha}|^2$ particles/sec. Since every intermediate state must disintegrate into one of the available channels,

$$\Sigma_\beta |S_{\beta\alpha}|^2 = 1 \qquad 13.6$$

If the same intermediate state is formed by a simultaneous flux of particles in channels α, β, described by two orthogonal wave functions

$$\Psi_\alpha, \Psi_\beta, \quad (\Psi_\alpha|\Psi_\beta) = 0, \quad \alpha \neq \beta,$$

the outgoing waves in all channels must also be orthogonal

$$\Sigma_\gamma \; S^*_{\beta\gamma} \; S_{\alpha\gamma} = 0 \qquad 13.7$$

13.6 and 13.7 can be condensed in

$$\Sigma_\gamma \; S^*_{\beta\gamma} \; S_{\alpha\gamma} = \delta_{\beta\alpha} \quad \text{or} \quad SS^\dagger = 1 \qquad 13.8$$

The scattering matrix is unitary:

$$S^\dagger_{ij} = S^*_{ji} \qquad 13.9$$

This poses $\frac{1}{2}N(N+1)$ constraints on the $2N^2$ real parameters contained in the $N \times N$ S-matrix. Time reversal invariance requires the S-matrix to be symmetrical (no spin)

$$S_{\beta\alpha} = S_{\alpha\beta}; \quad S^*S = 1 \qquad 13.10$$

if the spin is taken into account (reciprocity).

$$S_{\beta(m)\alpha(m')} = S_{\alpha(-m')\beta(-m)} \qquad 13.11$$

For further details see Blatt & Weisskopf, p. 528.

Simple examples: For the case of one channel, the S-matrix has only one element $S_\ell = e^{2i\delta_\ell}$. From 13.5 we obtain

$$\sigma_\ell = \frac{4\pi}{k^2}(2\ell+1) \sin^2 \delta_\ell \qquad 13.12$$

as well known. In the case of two channels:

$$S_\ell = \begin{pmatrix} \eta_\ell^{\alpha\alpha} e^{2i\delta_\ell^{\alpha\alpha}} & \eta_\ell^{\beta\alpha} e^{2i\delta_\ell^{\beta\alpha}} \\ \eta_\ell^{\beta\alpha} e^{2i\delta_\ell^{\beta\alpha}} & \eta_\ell^{\beta\beta} e^{2i\delta_\ell^{\beta\beta}} \end{pmatrix} \qquad 13.13$$

with all δ's and η's <u>real</u>. This will give cross sections:

$$\sigma_\ell^{(\alpha\to\alpha)} = (2\ell+1)\frac{\pi}{k_\alpha^2}\left|\eta_\ell^{\alpha\alpha} e^{2i\delta_\ell^{\alpha\alpha}} - 1\right|^2 \qquad 13.14$$

which is equivalent to our previous result 10.21, p. 103.

$$\sigma_\ell^{(\alpha\to\beta)} = (2\ell+1)\frac{\pi}{k_\alpha^2}\left|\eta_\ell^{\beta\alpha}\right|^2 \qquad 13.15$$

By using 13.6 we have

$$\left|\eta_\ell^{\alpha\alpha}\right|^2 + \left|\eta_\ell^{\beta\alpha}\right|^2 = 1 \quad \text{or} \quad \left|\eta_\ell^{\beta\alpha}\right|^2 = 1 - \left|\eta_\ell^{\alpha\alpha}\right|^2 \qquad 13.16$$

thus

$$\sigma_\ell^{(\alpha\to\beta)} = (2\ell+1)\frac{\pi}{k_\alpha^2}\left(1 - \left|\eta_\ell^{\alpha\alpha}\right|^2\right) \qquad 13.17$$

in the form equivalent to 10.22, p. 103.

We have already used the notation T_ℓ to denote the partial wave amplitudes. In our notation so far we can define a transition matrix T from

$$S = \mathbb{1} + 2iT \qquad 13.18$$

where $\mathbb{1}$ is the unit matrix. We then reobtain

$$T_\ell^{\alpha\alpha} = \frac{S-1}{2i} = \frac{\eta_\ell^{\alpha\alpha} e^{2i\delta_\ell^{\alpha\alpha}}}{2i} \qquad 13.19$$

in the previous form for the elastic scattering partial wave amplitude.

The reaction amplitude becomes:

$$T_\ell^{\alpha\beta} = \frac{S}{2i} = \frac{T_\ell^{\alpha\beta} e^{2i\delta_\ell^{\alpha\beta}}}{2i} \qquad 13.20$$

Argand Diagrams of the Elastic and Inelastic T_ℓ Amplitudes

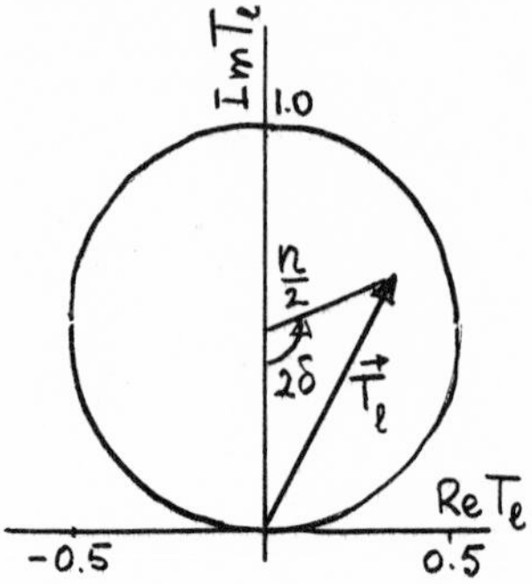

For a given ℓ or J, dropping the subscripts and superscripts, we have from 13.19 that

$$\begin{aligned} \text{Re } T_\ell &= \frac{\eta}{2} \sin 2\delta \\ \text{Im } T_\ell &= \frac{1}{2}(1 - \eta \cos 2\delta) \end{aligned} \qquad 13.21$$

In the complex plane, T_ℓ is a vector that lies on (for $\eta = 1$) or inside a circle of radius $\frac{1}{2}$ and center on the imaginary axis at $i/2$. The circle is called the unitary circle. The fact that the circle lies on the positive part of the imaginary axis is due to the fact that $\text{Im}T_{\alpha\alpha} = \sum_\gamma |T_{\gamma\alpha}|^2 > 0$ which follows from 13.8, 13.18, and time reversal invariance.

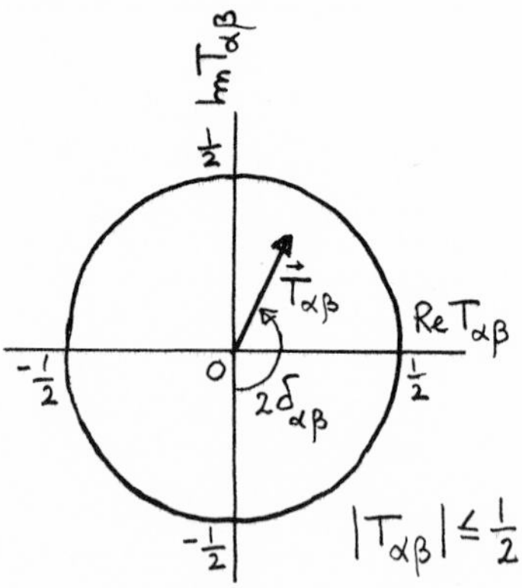

From 13.20 we have instead:

$$\begin{aligned} \text{Re } T_{\alpha\beta} &= \frac{1}{2} \eta_{\alpha\beta} \sin 2\delta_{\alpha\beta} \\ \text{Im } T_{\alpha\beta} &= -\frac{1}{2}\eta_{\alpha\beta} \cos 2\delta_{\alpha\beta} \end{aligned} \qquad 13.22$$

The vector representing $T_{\alpha\beta}$ is contained within a circle of radius $\frac{1}{2}$, with center at the origin. For $\delta = \frac{\pi}{2}$, the amplitude is pure imaginary.

Important: unitarity imposes a special condition on $\text{Im } T_{\alpha\beta}$:

From $SS^\dagger = 1 \rightarrow \text{Im } T = T^\dagger T$; $\quad \text{Im } T_{\beta\alpha} = \sum_n T^*_{n\beta} T_{n\alpha} \qquad 13.23$

for <u>each</u> partial wave.

For completeness, note that an alternative definition of the T matrix is sometimes used:

$$S_{\beta\alpha} = \delta_{\beta\alpha} + 2i\, k_\alpha^{1/2}\, k_\beta^{1/2}\, T_{\beta\alpha} \qquad 13.23$$

or

$$S = \mathbb{1} + 2i\, K^{1/2}\, T\, K^{1/2} \rightarrow T = K^{-1/2}\, \frac{S-1}{2i}\, K^{-1/2} \qquad 13.24$$

where K is a diagonal matrix of the channel momenta. If this definition is adopted,

$$\sigma_{\ell\alpha}^{(\alpha\rightarrow\beta)} = 4\pi(2\ell_\alpha+1)\, \frac{k_\beta}{k_\alpha} \left| T_{\beta\alpha} \right|^2 \qquad 13.25$$

where for one channel

$$T_\ell = \frac{e^{2i\delta_\ell}-1}{2ik} \qquad 13.26$$

This differs from our previous definition by a k in the denominator. The wave function for the collision process, in terms of this definition of the T-matrix, has the asymptotic form $(kr \rightarrow \infty)$

$$\Psi_{\ell_\beta}^{(\alpha\rightarrow\beta)}(r) = \delta_{\beta\alpha}\, \frac{\sin\,(k_\beta r - \frac{\ell\pi}{2})}{k_\beta r} + T_{\beta\alpha}\, \frac{e^{i(k_\beta r - \frac{\ell\pi}{2})}}{r}, \qquad 13.27$$

which is clearly of the form

$\Psi = e^{ikz} + f(\theta)\, \dfrac{e^{ikr}}{r}$. 13.27 represents a wave of unit amplitude incident in the α-channel, and outgoing waves of amplitude $T_{\beta\alpha}$ in channel β. If there are standing waves in some of the channels, $T_{\beta\alpha}$ will be replaced by $K_{\beta\alpha}$, the reaction or reactance matrix K.

14

RESONANT BEHAVIOUR OF PARTIAL WAVE AMPLITUDES

Elastic and Reaction Resonant Amplitudes

The formation of baryon resonances is announced by the observation of resonant behaviour for a particular partial wave. This is often erroneously defined to correspond to the case of a particular phase shift δ_ℓ going through $\frac{\pi}{2}$ (we shall see that this is not always the case) and described by the Breit-Wigner formula.

There are many ways to introduce the latter. We can derive, for example, from 12.9, p. 121, which was said to have poles:

$$\tan \delta_\ell = \frac{(\ell+1) - f_\ell(k)}{\ell + f_\ell(k)} \frac{(kr_0)^{2\ell+1}}{(2\ell-1)!!(2\ell+1)!!} \qquad 14.1$$

In Sect. 12 we have assumed that for small kr_0, the quantity $\alpha_\ell = \frac{\ell + 1 - f_\ell(k)}{\ell + f_\ell(k)}$ is $\approx$ constant. This is obviously not true for

$$f_\ell(k) = -\ell \qquad 14.2$$

where α_ℓ has a pole. In this case $\tan \delta_\ell \to \infty$, $\delta_\ell \to \frac{\pi}{2}$ at a particular value of $k = k_R$. We can expand $f_\ell(k)$ for values of k close to k_R, or equivalently for values of the energy E close to

$$E_R = \hbar^2 k^2_{R/2\mu}$$

$$f_\ell(E) \simeq -\ell + (E_R - E)\left(\frac{df_\ell}{dE}\right)_{E=E_R} + \ldots \qquad 14.3$$

and insert this in 14.1.

$$\tan \delta_\ell \simeq \frac{\Gamma_{\ell/2}}{E_R - E} \qquad 14.4$$

where

$$\Gamma_{\ell/2} = \frac{(kr_0)^{2\ell+1}}{\left(\frac{df_\ell}{dE}\right)_{E_R} \left[(2\ell-1)!!\right]^{-2}} \qquad 14.5$$

and a monotonically increasing term

$$-(kr_0)^{2\ell+1} / (2\ell+1)!!(2\ell-1)!!$$

has been neglected. From

$$T_\ell = \frac{1}{\cot \delta_\ell - i}$$

and 14.4 we obtain

$$T_{\ell,R} = \frac{\Gamma_{\ell/2}}{(E_R - E) - i\Gamma_{\ell/2}} \qquad 14.6$$

which is the Breit-Wigner amplitude for elastic scattering. (With no background amplitude.) More generally:

$$T_{\alpha\beta} = \frac{\pm \frac{1}{2}\sqrt{\Gamma_\alpha \Gamma_\beta}}{(E_R - E) - i\frac{\Gamma}{2}} \qquad 14.7$$

where Γ_α and Γ_β are the partial widths for the incident channel α and the outgoing channel β, Γ is the total width

$$\Gamma = \Gamma_\alpha + \Gamma_\beta + \ldots \qquad 14.8$$

The $\pm$ sign is related to the SU(3) couplings as seen previously. It is customary to introduce the parametrization

$$\varepsilon = \frac{E_R - E}{\Gamma/2} \quad \text{and} \quad x_i = \Gamma_i/\Gamma \quad (i = \alpha, \beta)$$
$$x_\alpha = \Gamma_\alpha/\Gamma = \text{elasticity}. \qquad 14.9$$

In these terms 14.7 becomes

$$T_{\alpha\beta} = \pm \frac{\sqrt{x_\alpha x_\beta}}{\varepsilon - i} \qquad 14.10$$

For the elastic channel, $x_\alpha = x_\beta$, and

$$T_{\alpha\alpha} = \frac{x_\alpha}{\varepsilon - i} \qquad 14.11$$

In this form, the graphical representation of the resonant amplitude is particularly simple. From 14.10:

$$\mathrm{Re}\ T_{\alpha\beta} = \pm \frac{\varepsilon\sqrt{x_\alpha x_\beta}}{1+\varepsilon^2} \qquad \mathrm{Im}\ T_{\alpha\beta} = \frac{\sqrt{x_\alpha x_\beta}}{1+\varepsilon^2} \qquad 14.12$$

The points at $\varepsilon = \pm 1$ correspond to $E = E_R \pm \frac{\Gamma}{2}$.

$$|\mathrm{Re}\ T_{\alpha\beta}| = |\mathrm{Im}\ T_{\alpha\beta}| = \left|\frac{\sqrt{x_\alpha x_\beta}}{2}\right| \qquad 14.13$$

At resonance, for the elastic channel, the relation between η and the elasticity x_α is

$$\eta = |2x_\alpha - 1| \qquad 14.14$$

The resonant amplitude describes a circle in the complex plane, of which 14.12 are the parametric equations. The circle is described counter-clockwise, due to the Wigner causality condition. (See e.g., Williams, p. 116 (1971) edition.) In the case of one channel, $\eta = x_\alpha = 1$, the resonant amplitude describes a circle of diameter 1, which is the unitary circle.

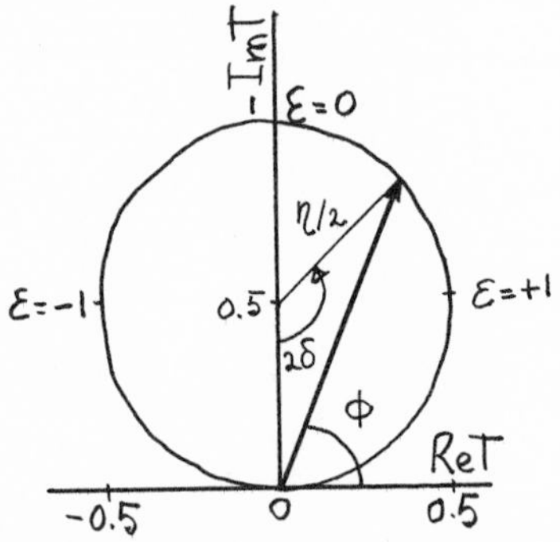

$$\mathrm{Re}\ T = \frac{\varepsilon}{1+\varepsilon^2} \qquad \mathrm{Im}\ T = \frac{1}{1+\varepsilon^2}$$

The phase angle defined by $\tan\phi = \frac{1}{\varepsilon}$ is called the resonant eigenphase. For $x_\alpha < 1$, the amplitude describes a smaller circle, of amplitude (diameter) x_α. At resonance the phase is $\delta = 90^\circ$ for $x \geq 0.5$, $\delta = 0^\circ$ for $x < 0.5$, but the eigenphase ϕ goes through 90° in both cases.

The Argand diagram of a resonant reaction amplitude is subject to the condition

$|T_{\alpha\beta}| \leq 1/2$ as seen before,

which becomes

$$\left|\sqrt{x_\alpha x_\beta}\right| \leq 1/2 \qquad 14.15$$

or $x_\alpha x_\beta < 0.25$

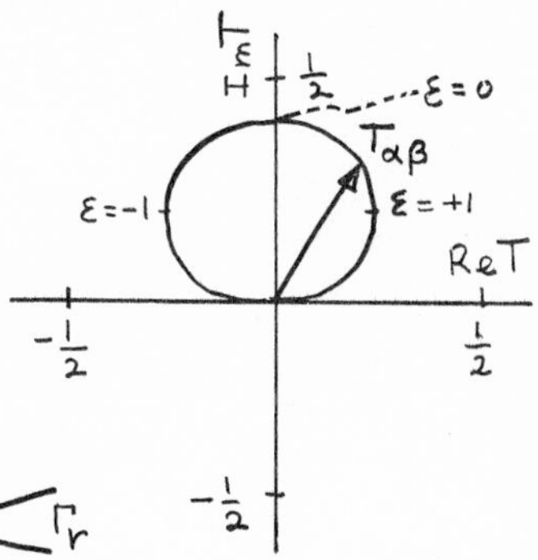

For a two-channel case, the T-matrix takes the form, calling $x = \frac{\Gamma_e}{\Gamma}$, since

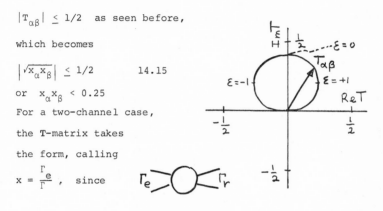

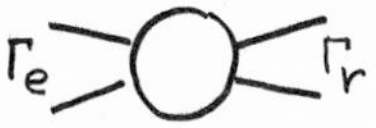

$$\frac{\Gamma_r}{\Gamma} = \frac{(\Gamma - \Gamma_e)}{\Gamma} = 1 - x \qquad 14.16$$

$$T_{\alpha\beta} = \frac{1}{\varepsilon - i}\begin{pmatrix} x & \sqrt{x(1-x)} \\ \sqrt{x(1-x)} & 1-x \end{pmatrix} \qquad 14.17$$

The S-matrix corresponding to this T-matrix, is unitary.

Another important property of the simple Breit-Wigner amplitude is, from 14.12

$$\left|\frac{dT}{d\varepsilon}\right| = \text{Im}T \qquad 14.18$$

for x = constant. For Γ = constant, ε is a linear function of energy and the maximum speed along the circle corresponds to the maximum of ImT, namely at the top of the circle. This property is used to determine the resonant energy of a partial wave amplitude, determined in an "energy independent" manner.

Energy Dependence of the Resonance Widths

The decay width of a resonant state depends in general on the energy and orbital angular momentum ℓ. We have in effect already derived an expression for such energy dependence, in a particular case of small kR, in, e.g., 14.5, valid for $kr \to 0$, or $kr \ll \ell$, or "very large ℓ." This expression can be related to the nomenclature in Blatt and Weisskopf, p. 410 and 361: the channel width is factorized as

$$\frac{\Gamma_\ell}{2} = \frac{(kr_0)^{2\ell+1}}{\left(\frac{df_\ell}{dE}\right)_{E_R}\left[(2\ell-1)!!\right]^2} \qquad 14.5$$

$$\Gamma_\alpha = 2k_\alpha r\, v_{\ell\alpha}\, \gamma_\alpha \qquad 14.19$$

where $\gamma_\alpha = \left[\left(\frac{df_\ell}{dE}\right)_{E_R}\right]^{-1}$ = reduced width, $v_{\ell\alpha}$ = barrier penetration factor $= \frac{(kr)^{2\ell}}{D_\ell}$ where D_ℓ is a function of kr and ℓ.

The additional factor kr in 14.19 can be regarded as a phase space factor. The function $v_{\ell\alpha}$ is just $B_\ell(k)$ of 8.2 on page 76 . In general then

$$\Gamma = \gamma \cdot \frac{(kr)^{2\ell+1}}{D_\ell} \qquad 14.20$$

These expressions have been derived for a square well potential and non-relativistic energies. A relativistic correction has been suggested by Layson in the form of an additional factor $2E_R/(E_R+E)$, normalized in such a way as to be equal to one at the resonant energy. The final expression for Γ can be taken as

$$\Gamma(E) = \Gamma_R \frac{(kR)_E^{2\ell+1}}{D_\ell(E)} \frac{2E_R}{E_R+E} \frac{D_\ell(R)}{(kR)_R^{2\ell+1}} \qquad 14.21$$

where the $D_\ell(E)$ and $D_\ell(R)$ are the D_ℓ's calculated for values of k corresponding to E and $E = E_R$ respectively. The radius of interaction R is an important parameter affecting the shape of the amplitude as a function of energy. When $\ell \neq 0$ ImT and ReT develop a tail at the high energy side:

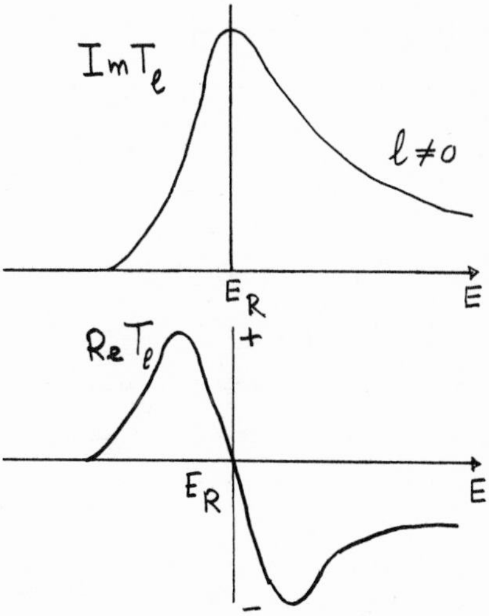

Effect of Resonances on Cross Sections

Reminder about statistical factor: consider the collision of two particles of spin S_1, S_2, to form a resonant state of total angular momentum J, for a given ℓ.

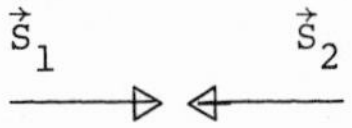

$\vec{S} = \vec{S}_1 + \vec{S}_2$ = "channel spin"

Number of possible states of channel spin: $(2S_1+1)(2S_2+1)$. For unpolarized beam: $2S + 1$ states for every value of channel spin S. The relative probability that the incident particle and the target particle will form a state of spin S is:

$$g(S) = \frac{2S+1}{(2S_1+1)(2S_2+1)} \qquad 14.22$$

Similarly, the relative probability of a state J out of a given channel spin S and orbital angular momentum ℓ is

$$P(J) = \frac{2J+1}{(2\ \ +1)(2S+1)} \qquad 14.23$$

The overall probability of formation of the state in question is then:

$$P(J,\ell,S) = (2\ell+1)g(S)P(J) = \frac{2J+1}{(2S_1+1)(2S_2+1)} \qquad 14.24$$

In our case $(0^- + \frac{1}{2}^+)$, 14.24 reduces to $J + \frac{1}{2}$. For the cross sections then:

$$\sigma_{\alpha\beta} = 4\pi\bar{\lambda}^2 (J+\tfrac{1}{2}) \left|T_{\alpha\beta}\right|^2 \qquad 14.25$$

which, for the Breit-Wigner amplitudes described by 14.10 yields, for the formation of a state IJP, and in absence of background contributions to the same state

$$\sigma^R_{\alpha\alpha} = 4\pi\bar{\lambda}\ (J+\tfrac{1}{2})\ \frac{x_\alpha^2}{1+\varepsilon^2}$$

$$\sigma^R_{\alpha\beta} = 4\pi\bar{\lambda}^2 (J+\tfrac{1}{2})\ \frac{x_\alpha x_\beta}{1+\varepsilon^2} \qquad 14.26$$

and for the total cross section

$$\sigma^R_{tot} = 4\pi\bar{\lambda}^2 (J+\tfrac{1}{2})\ \frac{x_\alpha}{1+\varepsilon^2} \qquad 14.27$$

Note first of all that the elastic cross section is proportional to x_α^2, while the total cross section is proportional to x_α.

Thus the comparison of the size of the resonant bump in σ_{el} and σ_{tot} can give an indication of the spin J of the resonance. Second, since x_α is in general < 1, resonances show up as more prominent bumps in σ_{tot} than in σ_{el}. To be noted also is the fact that the factor $\lambdabar^2$ has the effect of displacing the position of the peak in the cross sections, when compared with that in the amplitude. The displacement of the peak depends on the slope of $4\pi\lambdabar^2$ in the resonance region, and also on the width of the resonance.

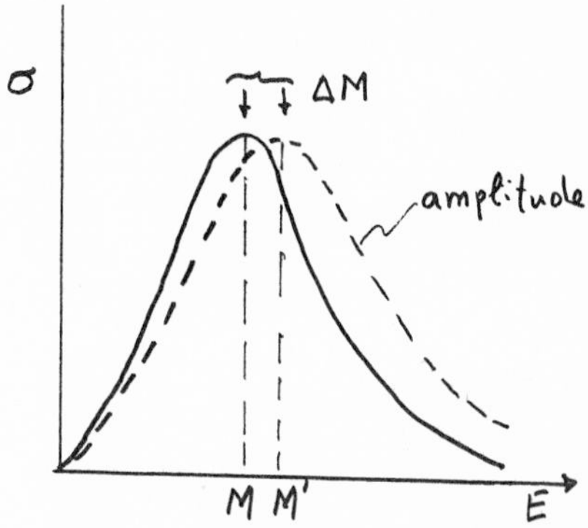

For this reason the exp. data are often studied in plots of $\sigma/4\pi\lambdabar^2$, to eliminate the distortion due to the kinematical factor.

Interfering Resonances

Resonances in meson-baryon systems are seldom isolated and important interference effects occur frequently. Such effects can be very useful in the determination of spin-parity of resonant amplitudes. We have seen that the differential cross sections and polarization in $0^- - \frac{1}{2}^+$ scattering are often expanded in Legendre polynomial series:

$$\frac{d\sigma}{d\Omega} = \lambdabar^2 \Sigma_n A_n P_n(\mu)$$

$$\frac{d\sigma}{d\Omega}\vec{p}(\theta) = \hat{n}\lambdabar^2 \Sigma_n B_n P_n^1(\mu) \qquad 14.28$$

where $\mu = \cos\theta$. The coefficients A_n and B_n are of the form:

$$A_n = \sum_{i \leq j} a_{ij} \mathrm{Re}(T_i^* T_j) \qquad 14.29$$

as seen in the tabulation on pages 116, 117.

$$B_n = \sum_{i \leq j} b_{ij} \text{ Im } (T_i^* T_j) \qquad 14.30$$

Whenever two amplitudes T_i, T_j are both resonant, the coefficients of the Legendre expansion will have the following behavior:

$$A_n \sim \text{Re } (T_i^* T_j) \sim \frac{x_1 x_2 (1+\varepsilon_1 \varepsilon_2)}{(1+\varepsilon_1^2)(1+\varepsilon_2)^2} \qquad 14.31$$

$$B_n \sim \text{Im } (T_i^* T_j) \sim \frac{x_1 x_2 (\varepsilon_1 - \varepsilon_2)}{(1+\varepsilon_1^2)(1+\varepsilon_2^2)} \qquad 14.32$$

If the resonances are widely spaced, at least one of the amplitudes is predominantly real over the region of interference. The resonance energies are then close to the nodes of the interference pattern. Note also that the sign of the interference, for inelastic channels, depends on the product of the relative couplings to the resonant states.

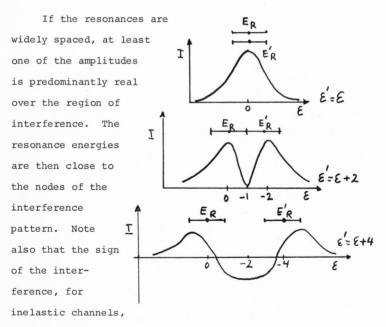

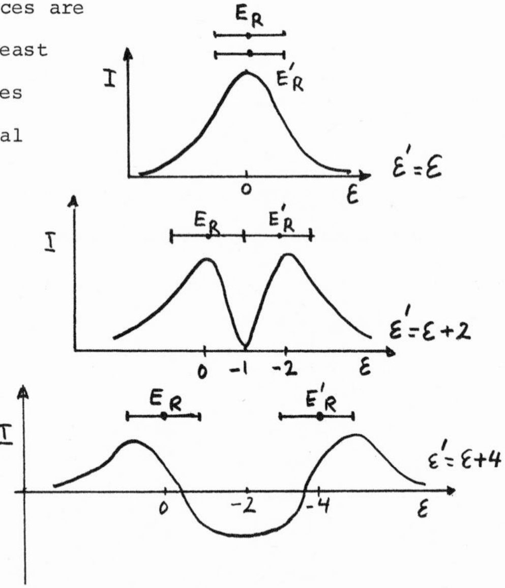

Effect of Background on Resonant Amplitude

So far the resonant partial wave amplitude has been considered as representing a pure state, characterized by a set of quantum numbers I,J,P. Often, for the lower partial waves in particular, in addition to the resonant state, also a non-resonant amplitude contributes to the overall partial wave amplitude. If the non-resonant part is elastic, we have a situation frequently encountered in nuclear physics, where "potential scattering" or "hard-sphere" scattering may contribute and interfere with "resonant scattering." In combining the background and resonant amplitudes, attention should be paid to the requirement of unitarity. The two amplitudes cannot just be added, since the overall amplitude can then fall outside the unitary circle. A way to

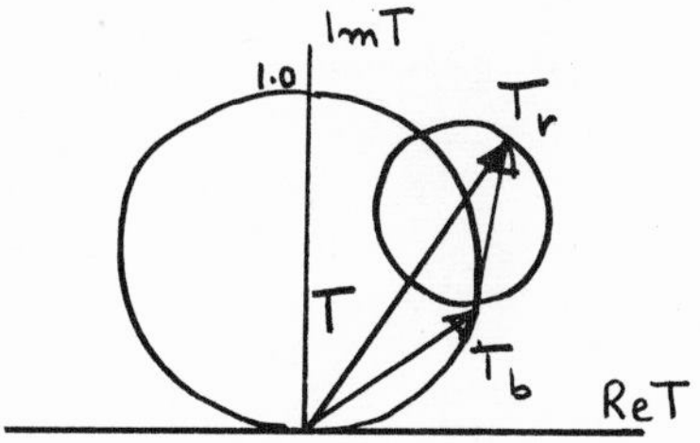

ensure that unitarity is satisfied is to write the S matrix in factorable form

$$S = S_r S_b \tag{14.33}$$

If S_r and S_b are each unitary, then S is also unitary.

Although physically such a factorization would be plausible if the background and resonant interactions were to take place in distinct radial regions (a rather unlikely situation) this approach is widely adopted with a degree of success.

From 14.33 we obtain

$$S = 1 + 2iT = (1 + 2iT_r)(1 + 2iT_b) \qquad 14.34$$

from which

$$T = T_b + T_r S_b \qquad 14.35$$

If the background is purely elastic, which is often the case,

$$S_b = e^{2i\delta_b}; \qquad T = T_b + e^{2i\delta_b} T_r \qquad 14.36$$

This result can be obtained in a number of other ways. For example, following a derivation by Dalitz, we can write the partial wave amplitude which exhibits resonant behavior as

$$T = e^{i\delta_r} \sin \delta_r = \frac{e^{2i(\delta_\infty + \delta)} - 1}{2i} = \frac{e^{2i\delta_\infty}(e^{2i\delta} - 1)}{2i} + \frac{e^{2i\delta_\infty} - 1}{2i} \qquad 14.37$$

Here $\delta_r = \delta_\infty + \delta$ where δ_∞ is to be regarded as the phase shift far away from the resonance. Now, the usual expansion of

$$\frac{e^{2i\delta} - 1}{2i} = \frac{\gamma}{E_R - E - i\gamma}$$

yields

$$e^{i\delta_r} \sin \delta_r = e^{2i\delta_\infty} \frac{\gamma}{E_R - E - i\gamma} + e^{i\delta_\infty} \sin \delta_\infty \qquad 14.38$$

which is clearly of the form 14.36. This applies to a one-channel situation.

A plot of 14.36 or 14.38, in the case of one channel, is a resonant circle, rotated by $2\delta_\infty$

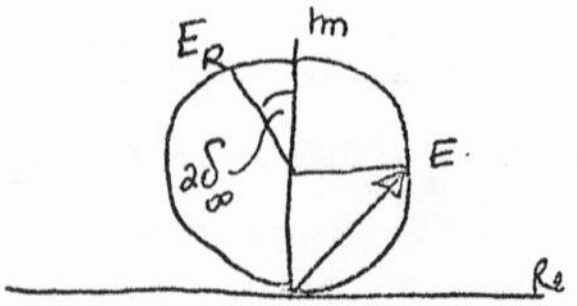

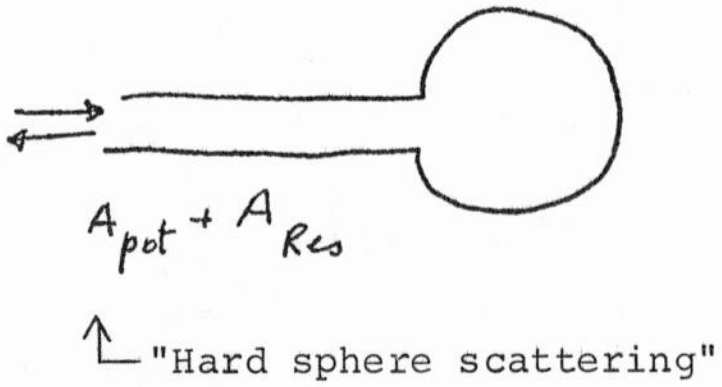

The above considerations can be extended to the case in which the resonant amplitude may involve more than one channel (inelastic resonance), but still an elastic background term, by writing

$$T_{\alpha\alpha} = e^{i\delta_b} \sin \delta_b + e^{2i\delta_b} \frac{x}{\varepsilon - i} \qquad 14.39$$

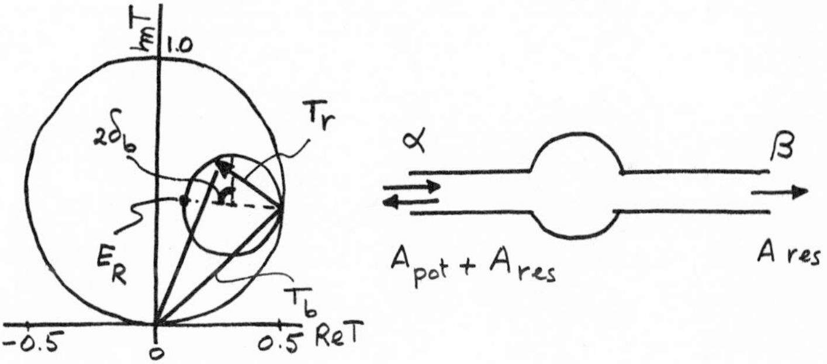

In the more general case in which also the background amplitude may be inelastic, the total elastic amplitude takes the form

$$T_{\alpha\alpha} = \frac{\eta_b e^{2i\delta_b} - 1}{2i} + e^{2i\delta_\infty} \frac{x}{\varepsilon - i} \qquad 14.40$$

where now the phase δ_∞ is different from δ_b, the latter containing a contribution from all the background channels including the elastic. The resonant circle is rotated by $2\delta_\infty$ with respect to the case of a pure imaginary amplitude. Goebel and McVoy (PR164,1932,1967) refer to δ_∞ as a relative phase between background and resonance. While in general there is one such phase for each channel ($\alpha\beta$), unitary relates these phases to one another and the background and resonance parameters. In the case of a two channel problem, Coulter (PRL 29,450,1972) has explicitly determined these phases in terms of the background η and the channel widths of the resonance.

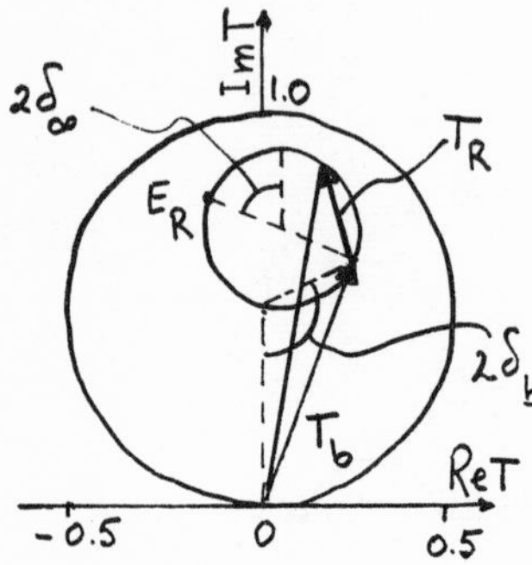

A circular trajectory of the amplitude $T_{\alpha\alpha}$ clearly corresponds to the idealized case of a background amplitude constant with energy. Since this is usually not the case, the actual trajectory of a resonant amplitude in the presence of background may be quite distorted.

For example:

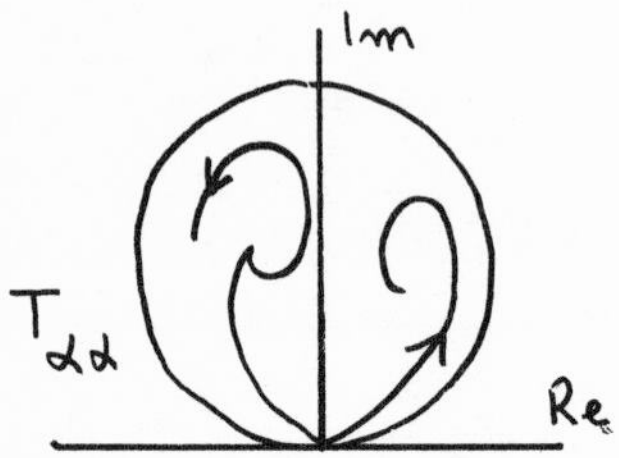

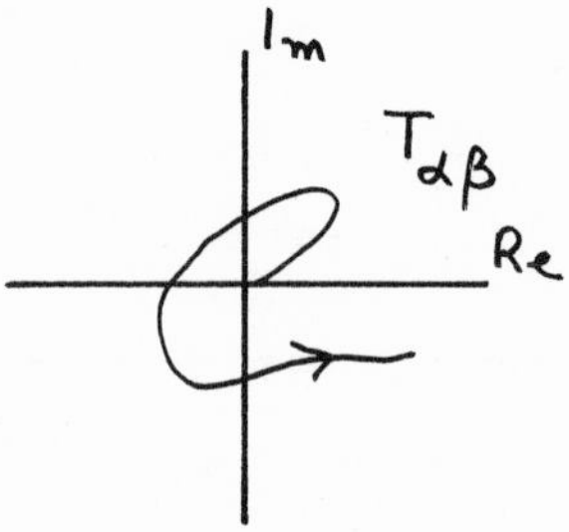

References

Much of the material in lectures 10-14 covers well-known topics in quantum-mechanics and nuclear reaction theory. Basic reference textbooks for the latter are:

N.F. Mott and H.S.W. Massey, Theory of Atomic Collisions, Oxford University Press, London, 1949.

J.M. Blatt and V.F. Weisskopf, Theoretical Nuclear Physics, Wiley, New York, 1952.

L.D. Landau and E.M. Lifshitz, Quantum Mechanics, Pergamon, New York, 1958.

E. Segrè, Nuclei and Particles, Benjamin, New York, 1964.

Particular presentations originate from exposure to lecture material by:

R.D. Tripp, Baryon Resonances, Lectures at the Varenna International School of Physics on Strong Interactions, July 1964 and the 1965 Easter School at Bad Kreuznach, April 1965. CERN Report 65-7 (Rev.).

M. Ferro-Luzzi, Three Lectures on Baryon Resonances, Proceedings of the 1968 CERN School of Physics, El Escorial, May 1968, CERN Report 68-23.

G. Costa, Topics on elementary theory of scattering. Proceedings of the 1964 Easter School, Herceg-Novi, May 1964. CERN Report 64-13, Vol. III.

An excellent compendium of the topics discussed here, and from which inspiration has been liberally drawn, is the review by A. Barbaro-Galtieri, Baryon Resonances, Advances in Particle Physics, Vol. 2. Edited by R.L. Cool and R.E. Marshak, Interscience, 1968.

Other satisfying sources are:

R.H. Dalitz, Strange-Particle Resonant States, in Annual Review of Nuclear Science 13, 339, 1963.

D.H. Miller. The Elementary Particles with Strong Interactions, in High Energy Physics, Vol. IV, edited by E.H.S. Burhop, Academic Press, New York and London, 1969.

W.S.C. Williams, An Introduction to Elementary Particles, Second Edition. Academic Press, New York, London, 1971.

15

FURTHER TOPICS IN SCATTERING THEORY: UNITARITY FROM THE "S-MATRIX" POINT OF VIEW

The scattering formalism developed in the preceding lectures assumed the existence of a potential theory and a corresponding (Schrödinger) wave equation. In this and the following few lectures we approach the same problem from the so-called "analytic S-matrix" point of view. We assume the existence of a matrix S which connects initial and final states i and f. In contrast to the S-matrix introduced in earlier lectures, this matrix is for the full scattering amplitude, not for its partial wave projection. In particular f corresponds to any number of final state particles. From this rather general assumption we shall derive and extend the matrix formalism for coupled channels discussed earlier. In addition we shall see how resonances can arise in the formalism without having been "a priori" assumed.

The aim of this Lecture is primarily to define notation and to introduce the concept of partial wave unitarity for two body final states. Consider the coupled channel problem of $\pi\pi \to \pi\pi$, $\bar{K}K \to \bar{K}K$, and $\pi\pi \to \bar{K}K$. We shall show, starting from rather general assumptions, that for each partial wave the S-matrix may be written as

$$\begin{pmatrix} S_{11} & S_{12} \\ S_{12} & S_{22} \end{pmatrix} = \begin{pmatrix} 1 & 0 \\ 0 & 1 \end{pmatrix} + 2i \begin{pmatrix} q_1^{1/2} & 0 \\ 0 & q_2^{1/2} \end{pmatrix} \begin{pmatrix} T_{11} & T_{12} \\ T_{12} & T_{22} \end{pmatrix} \begin{pmatrix} q_1^{1/2} & 0 \\ 0 & q_2^{1/2} \end{pmatrix} \qquad 15.1$$

where

$$\frac{d\sigma_{ij}}{d\Omega} = \frac{q_i}{q_i} |T_{ij}|^2 \qquad 15.2$$

for $i,j = 1,2$ corresponding to the $\pi\pi, \bar{K}K$ channels and q_i, the momenta in those channels. Since S will be unitary it can in this case be written as

$$S = \begin{pmatrix} \eta e^{2i\delta_1} & i\sqrt{1-\eta^2}\, e^{i\delta_1+i\delta_2} \\ i\sqrt{1-\eta^2}\, e^{i\delta_1+i\delta_2} & \eta e^{2i\delta_2} \end{pmatrix}$$

where $\delta_1(\pi\pi)$ and $\delta_2(\bar{K}K)$ are arbitrary real phases and $0 \leq \eta \leq 1$.

Cross Sections and the Optical Theorem

We assume that the cross section for scattering from state i to state f is given by

$$\sigma_{ij} = \frac{1}{4F}\int |M_{ij}|^2 (2\pi) 4\delta^4(p_i - p_f) \frac{d^3\vec{q}_1}{2E_1(2\pi)^3} \cdots \frac{d^3\vec{q}_n}{2E_n(2\pi)^3} \qquad 15.3$$

where $p_i = p_2$, $p_f = q_1 + \ldots + q_n$ (n particles in final state)

$$F = \{(p_1p_2)^2 - p_1^2 p_2^2\}^{1/2} = p_{cm}\sqrt{s} = p_{lab} m_{target}$$

E_i = total energies of final particles.

In general the labels i,j denote all physically observable attributes such as particle label, spin, momentum, etc.

The matrix elements M_{ij} are related to the S matrix by

$$S_{if} = \delta_{if} + i(2\pi)^4 \delta^4(P_i - P_f) M_{if} \qquad 15.4$$

(See H. Pilkuhn, Chapter 182, for the derivation of equations 15.3 and 15.4.) An immediate consequence of the unitary property of S, $\sum_i S_{ij}(S^\dagger)_{jf} = \delta_{if}$, is the optical theorem. Using equation 15.4 we have

$$i(2\pi)^4 \sum_j \{\delta_{jf}\, \delta^4(P_i - P_j) M_{ij} - \delta_{ij} \delta^4(P_f - P_j) M^*_{fj}\}$$

$$= + [i(2\pi)^4]^2 \sum_j M_{ij} M^*_{fj}\, \delta^4(P_i - P_j) \delta^4(P_f - P_j)$$

Since formally $\delta^{(4)}(P_i - P_f)\delta^{(4)}(P_f - P_j) = \delta^{(4)}(P_i - P_j)\delta^{(4)}(P_i - P_f)$,

$$M_{if} - M^*_{fi} = i(2\pi)^4 \sum_j \delta^4(P_i - P_f) M_{ij} M^*_{fj} \qquad 15.5$$

Comment: Note that $\sum_j$ denotes a sum over all possible physical intermediate states,

$$\sum_j = \sum_{\text{channels}} \sum_{\text{spin}} \prod_{i=1}^{j} \int \frac{d^3 q_i}{2E_i (2E)^3} \cdots$$

Let i and f denote the same two particle state; that is, M_{ii} is the elastic amplitude in the forward direction (t=0). Then

$$\text{Im } M_{ii}(s, t=0) = \frac{1}{2} \sum_j \prod_{k=2}^{N} \left(\int \frac{d^3 q_k}{(2E)^3 2E_k} \right) |M_{ij}|^2 (2\pi)^4 \delta^4(P_i - P_j)$$

From the definition of cross section, equation 15.3, we have

$$\text{Im } M_{ii}(s,t=0) = 2p_{cm}\sqrt{s}\sum_j \sigma_{ij} = 2p_{cm}\sqrt{s}\;\sigma_{tot} \qquad 15.6$$

Comments: 1) If M_{ij} represented spin $0\text{-}1/2 \rightarrow 0\text{-}1/2$ scattering, then a trace of the spin matrix, is implied on the left hand side of equation 15.6. In this case we would have to divide Equation 15.5 by a factor of 2 (average over spin!) before we could equate it to $2p_{cm}\sqrt{s}\;\sigma_{tot}$.

2) We have left out the one particle intermediate state ($\pi p \rightarrow N$). Because of $\delta^{(4)}(p_i-p_j)$ this state gives a contribution only at the nucleon mass and does not affect Equation 15.5 for $s \neq M_N^2$. On the other hand, were we to consider a dispersion relation such as $\text{Re}M_{ii} = \frac{1}{\pi}\int_{-\infty}^{+\infty}\frac{\text{Im}M_{ii}}{s'-s}ds'$, the one particle state would give rise to the pole term which affects $\text{Re}M_{ii}$ for all s.

Two-Body Final States and Two-Body Unitarity

Consider the process (i) $q_1 + q_2 \rightarrow$ (f) $q_1' + q_2'$ where q_i is the cm momentum for the initial system and q_f, for the final system. Then

$$\sigma_{if} = \frac{1}{4F}\int |M_{if}|^2 (2\pi)^4 \delta^4(q_1+q_2-q_1'-q_2') \frac{d^3q_1'}{2E_1'(2\pi)^3} \frac{d^3q_2'}{2E_2'(2\pi)^3}$$

$$= \frac{1}{4q_i\sqrt{s}} \int |M_{if}|^2 \frac{\delta(\sqrt{s}-E_1'-E_2')q'^2 dq' d\Omega}{4E_1'E_2'(2\pi)^2}$$

$$= \frac{1}{(8\pi)^2 q_i\sqrt{s}} \quad |M_{if}|^2 \frac{\delta(q'-q_f)q'^2 dq' d\Omega}{E_1'E_2' \frac{q'\sqrt{s}}{E_1'E_2'}}$$

where we have used

$$\delta(\sqrt{s} - E_1' - E_2') = \frac{(q' - q_f)}{\left|\frac{q'}{E_1'} + \frac{q'}{E_2'}\right|}$$

Thus,

$$\frac{d\sigma_{if}}{d\Omega} = \frac{q_f}{q_i}\left|\frac{M_{if}}{8\pi\sqrt{s}}\right|^2 \overset{\text{def}}{\equiv} \frac{q_f}{q_i}|T_{if}|^2 \qquad 15.7$$

Equation 15.7 defines the amplitude T_{if} which we shall use below.

We shall see later that for q_f sufficiently small, $T_{if} \approx \frac{\beta}{1-iq_i\gamma}$ where β and γ are real constants. Thus we see that inelastic two body cross sections grow like q_f near their threshold.

We now study the properties of unitarity, equation 15.5, keeping only the two body intermediate states in Σ_j.

$$M_{if} - M_{fi}^* = i(2\pi)^4 \sum_{\substack{j=\text{channel,}\\ \text{spin}}} \int \delta^4(p_i-p_f)M_{ij}M_{fi}^* \frac{d^3q_1 d^3q_2}{4E,E_1(2\pi)^6},$$

where by channel we mean $\pi\pi \to \begin{cases} \pi\pi \\ KK \\ \vdots \end{cases}$.

Using $T_{ij} = \dfrac{M_{ij}}{8\pi\sqrt{s}}$ and disposing of $\delta^{(4)}(p_j - p_f)$ as in the derivation of Equation 15.7, we have

$$T_{if}(s,\theta) - T^*_{fi}(s,\theta) = 2i \sum_j q_i \int \frac{d\cos\theta' d\phi'}{4\pi} T^*_{ji}(s,\theta') T_{jf}(s,\theta''), \quad 15.8$$

where

$$\cos\theta'' = \cos\theta \cos\theta' + \sin\theta \sin\theta' \cos(\phi - \phi')$$

as illustrated.

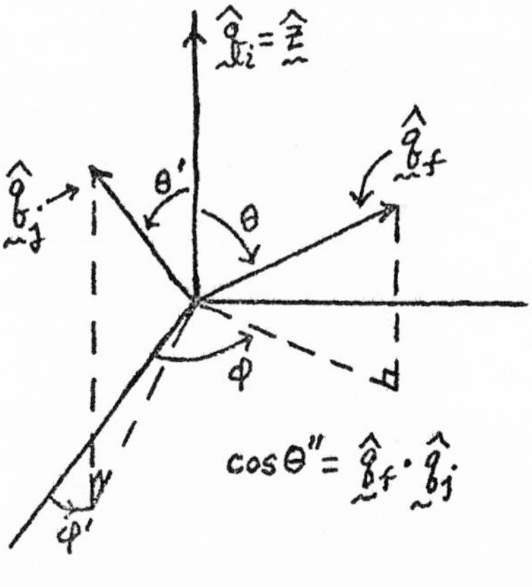

$$\hat{q}_f = \sin\theta \cos\phi \hat{x} + \sin\theta \sin\phi \hat{y} + \cos\theta \hat{z}$$

$$\hat{q}_j = \sin\theta' \cos\phi' \hat{x} + \sin\theta' \sin\phi' \hat{y} + \cos\theta' \hat{z}$$

$$\cos\theta'' = \hat{q}_f \cdot \hat{q}_j$$

We next introduce the partial wave expansion for $T_{if}(s,\theta)$,

$$T_{if}(s,\theta) = (2\ell+1)\ T_{if}^{(\ell)}(s)\ P_{\ell}(\cos\theta).$$

Notice $T_{if}^{(\ell)}(s)$ is independent of angle. Below we set $z = \cos\theta$. Equation 15.9 is completely general. We choose to expand in $P_{\ell}(z)$'s because then the integral in Equation 15.8 is trivial. The end result of all this manipulation is that unitarity will be reduced to a set of algebraic equations.

Substituting Equation 15.8 into Equation 15.9, we have (note T_{if} is symmetric provided time reversal invariance holds!)

$$\sum_{\ell}(2\ell+1)\,\mathrm{Im}T_{if}^{(\ell)}(s)P_{\ell}(z)$$

$$= \sum_{j} q_i \int_{-1}^{+1}\frac{d\cos\theta'}{2}\int_{0}^{2\pi}\frac{d\phi'}{2\pi} \times \left\{\sum_{\ell'}(2\ell'+1)T_{ji}^{(\ell')*}(s)P_{\ell'}(z')\right\}$$

$$\times \left\{\sum_{\ell''}(2\ell''+1)T_{jf}^{(\ell)}(s)P_{\ell''}(z'')\right\}.$$

Using the relations

$$\int d\phi' P_{\ell}(z'') = 2\pi P_{\ell}(z)P_{\ell}(z')$$

and

$$\int_{-1}^{1} dz' P_{\ell}(Z')P_{\ell'}(z') = \frac{2\delta_{\ell\ell'}}{2\ell+1},$$

we have

$$\sum_{\ell}(2\ell+1)\,\mathrm{Im}T_{if}^{(\ell)}(s)P_{\ell}(z) = \sum_{j} q_j \sum_{\ell}(2\ell+1)T_{ji}^{*(\ell)}(s)T_{jf}^{(\ell)}(s)P_{\ell}(z)$$

or

$$\mathrm{Im}T_{if}^{(\ell)}(s) = \sum_{j} q_j T_{ji}^{(\ell)^{*}}(s)T_{jf}^{(\ell)}(s). \qquad 15.10$$

This result we shall refer to as partial wave unitarity for two bodies. The usefulness of Equation 15.10 is best illustrated by discussing how one would perform a multi-channel partial wave analysis. Consider the matrix

$$\begin{array}{cc} & \begin{array}{ccc} \bar{K}N & \pi\Lambda & \pi\Sigma \end{array} \\ T_{if}^{(\ell)}(s) = \begin{array}{c} \bar{K}N \\ \pi\Lambda \\ \pi\Sigma \end{array} & \begin{pmatrix} T_{11}(s) & T_{12}(s) & T_{13}(s) \\ & T_{22}(s) & T_{23}(s) \\ & & T_{33}(s) \end{pmatrix} \end{array}$$

One might start by parameterizing $T_{11}^{(\ell)}(s)$ and fitting Equation 15.9 to the elastic scattering data ($\bar{K}N \rightarrow \bar{K}N$). But what about the other available data: $\bar{K}N \rightarrow \pi\Sigma,\ \pi\Lambda$? Do we have to use a whole new set of parameters for each channel? Can unitarity, Equation 15.5, help us eliminate parameters? Unfortunately if you look at Equation 15.5, in particular at $\sum_j$, you find that if N body intermediate states are kept, a set of non-linear integral equations is obtained. The beauty of the two-body approximation is that we can reduce these integral equations to algebraic ones--Equation 15.10.

Let $Q^{1/2} = \begin{pmatrix} q_1^{1/2} & 0 \ldots & 0 \\ 0 & q_2^{1/2} & \vdots \\ \vdots & \ddots & \\ 0 & 0 \ldots & q_n^{1/2} \end{pmatrix}$, where the q_i

are the (n-) channel momenta. We may then define the partial-wave S-matrix, $S^{(\ell)}(s)$

$$S_{ij}^{(\ell)}(s) = \delta_{ij} + 2i(Q^{1/2}T^{(\ell)}(s)Q^{1/2})_{ij}$$
$$= \delta_{ij} + 2i\ Q_i^{1/2}\ Q_j^{1/2}\ T_{ij}^{(\ell)}(s). \qquad 15.11$$

We readily verify that $S^{(\ell)}$ is unitary.

$$I \overset{?}{=} S^{(\ell)\dagger}S^{(\ell)} = (I-2_iQ^{1/2}T^{(\ell)\dagger}Q^{1/2})(I+2iQ^{1/2}T^{(\ell)}Q^{1/2})$$
$$= I + 2iQ^{1/2}(T^{(\ell)}-T^{(\ell)\dagger})Q^{1/2}-(2i)^2Q^{1/2}T^{(\ell)\dagger}QT^{(\ell)}Q^{1/2}$$

Clearly, $S^{(\ell)}$ is unitary provided

$$T^{(\ell)} - T^{(\ell)\dagger} = 2iT^{(\ell)\dagger}QT^{(\ell)} \qquad 15.12$$

or $\quad \mathrm{Im}T^{(\ell)}(s) = T^{(\ell)\dagger}(s)\ QT\ (s)$

This is simply the matrix statement of Equation 15.10. From Equations 15.6 and 15.7 and $T = \dfrac{M}{q\sqrt{s}}$, we have

$$\sigma_{tot}^{(\ell)} = \frac{4\pi}{q_1}\ \mathrm{Im}T_{11}^{(\ell)}(2\ell+1) = \sigma_{11}^{(\ell)} + \sigma_{12}^{(\ell)}. \qquad 15.13$$

Comments:

1) Although the derivation presented here assumed that the scattered particles were spinless (cf. Equation 15.9),

the result, Equation 15.12, is valid for the more general case where the external particles have spin. Thus, for spin 0-1/2 scattering, the ℓ's in Equation 15.12 would become $\ell\pm$ corresponding to $J = \ell \pm 1/2$. This remark is most easily proven in the helicity representation. See, for example, H. Pilkuhn, The Interaction of Hadrons, §§ 3-8, 3-9.

2) The amplitude $T_{if}^{(\ell)}(s)$ is related to the partial wave or "Argand" amplitude $A_{if}^{(\ell)}(s)$ by

$$T_{if}^{(\ell)}(s) = \frac{1}{(q_i q_f)^{\frac{1}{2}}} A_{if}^{\ell}(s)$$

so that

$$T_{if}(s,\theta) = \sum_{\ell} (2\ell+1) \frac{A_{if}^{\ell}(s)}{(q_i q_f)^{\frac{1}{2}}} P_{\ell}(\cos\theta).$$

References:

General and K-Matrix:

R.H. Dalitz, Am. Rev. Nucl. Sci. 13, 339(1963); Rev. Mod. Phys. 33, 471(1961); Strange Particles and Strong Interactions, Oxford U. Press, 1962.

A.M. Lane and R.G. Thomas, Rev. Mod. Phys. 30, 257(1958)--An exhaustive review of the Reaction matrix formalism from the viewpoint of potential theory.

Two Body Unitarity:

H. Pilkuhn, The Interaction of Hadrons, John Wiley & Sons, 1967--extracted and modified from Chapters 1 and 2.

16

THE K-MATRIX

Motivation

There are several excellent reasons for introducing the K-matrix formalism. They all center about the essential result of two body unitarity, Equation 15.10,

$$\text{Im}\ T_{if}^{(\ell)}(s) = \sum_j q_j\ T_{ji}^{(\ell)*}(s)\ T_{if}^{(\ell)}(s)$$

In practice the experimentalist attempts to fit data to partial wave expansions. Equation 15.10 tells him that the partial wave amplitudes for <u>different</u> channels ($\bar{K}N \to \bar{K}N$, $\bar{K}N \to \pi\Sigma$) are related in a rather specific manner. If we choose to work with the amplitudes $T_{if}^{(\ell)}(s)$, the relations (Equation 15.10) are still rather complicated even though they are now algebraic in nature.

As we shall see, the introduction of a real, symmetric matrix $K^{(\ell)}$ such that

$$T^{(\ell)}(s) = K^{(\ell)}(s)(I-iQK^{(\ell)}(s))^{-1}$$

insures that Equation 15.10 is satisfied provided we parameterize $K^{(\ell)}$ as real and symmetric. Furthermore, since $K^{(\ell)}$ is real and symmetric, it can be diagonalized (we shall diagonalize $Q^{1/2}K^{(\ell)}Q^{1/2}$). The utility of this is that the identification

$$K^{(\ell)} \leftrightarrow \tan \delta_\ell / q$$

can be made with the eigenvalues of $K^{(\ell)}$. This identification in turn leads to physical interpretations associated with with elastic scattering:

i) $q^{2\ell+1} \cot\delta_\ell = \frac{1}{a_\ell} \leftrightarrow q^\ell \frac{1}{K^{(\ell)}} q^\ell = A_\ell^{-1}$

ii) When an <u>eigenphase</u> passes through 90°, we have a resonance (the usual interpretation of an elastic phase), but not necessarily at the energy where the eigenphase is 90°.

It must be emphasized that such features as scattering lengths and resonances arise in the K-matrix formalism through only the assumption of a Taylor expansion ("effective range") for K^{-1}.

<u>The K-Matrix and Unitarity</u>

In the following discussion we shall drop the index ℓ, as well as the heretofore implicit index for isospin. Consider the S matrix for N-channels (note that it is a square, $N \times N$, matrix). It is easy to see that S can always be represented by a Hermitian matrix H. Suppose $S = e^{iH}$. Then

$$S^\dagger S = e^{-iH^*} e^{iH} = e^{i(H-H^\dagger)} e^{i\frac{1}{2}[H^\dagger,H]}$$

(see Messiah, p. 442--careful, these are matrices!) Clearly if we assume $H^\dagger = H$, then S is unitary. Furthermore, if time reversal invariance holds, then $S = S^T$ (symmetric) implies that H is real.

We will introduce K through

$$T = K(1 - iQK)^{-1} \qquad 16.1$$

where as above Q is a diagonal matrix of channel momenta. (We'll often mean the identity matrix when we write 1!) Recall that

$$S = 1 + 2iQ^{1/2}\ T\ Q^{1/2}. \qquad 15.11$$

We can easily verify that T is unitary (i.e., satisfies Equation 15.10 or 15.12)

$$\text{Im}T = T - T^{\dagger} = T^{\dagger}\ Q\ T.$$

From Equation 16.1,

$$T^{\dagger} = (1 + iK^{\dagger}Q)^{-1}\ K^{\dagger} = (1 + iKQ)^{-1}\ K,$$

since $K^{\dagger} = K$ (K real and symmetric).

Thus
$$\begin{aligned} T - T^{\dagger} &= K(1-iQK)^{-1} - (1+iKQ)^{-1}\ K \\ &= (1+iKQ)^{-1}[(1+iKQ)K - K(1-iQK)](1-iQK)^{-1} \\ &= (1+iKQ)^{-1}(2iKQK)(1-iQK)^{-1} \\ &= 2i\ [(1+iKQ)^{-1}K]Q[K(1-iQK)^{-1}] \\ &= 2i\ T^{+}\ Q\ T \end{aligned}$$

Comment: This result holds provided $K^{\dagger} = K$. The stronger statement, $K^{T} = K$ and $K^{\dagger} = K$, follows provided time reversal invariance holds so that the partial wave S-matrix is symmetric.

Cross Sections and Complex Scattering Lengths

We next familiarize ourselves with the K-matrix formalism by computing elastic and inelastic (partial-wave!) cross sections. Consider a two channel problem with $Q = \begin{pmatrix} k & 0 \\ 0 & q \end{pmatrix}$, k and q being the respective channel

CM-momenta. Suppose, for the ℓth partial wave, that

$$K^{(\ell)} = Q^{\ell}\begin{pmatrix} \alpha & \beta \\ \beta & \gamma \end{pmatrix}Q^{\ell} = \begin{pmatrix} \alpha k^{2\ell} & \beta(kq)^{\ell} \\ \beta(kq)^{\ell} & \gamma q^{2\ell} \end{pmatrix}, \qquad 16.2$$

where α,β,γ are constants. This choice of K will become clear once we compute T_{11} and through an identification with pure elastic scattering where β would be zero. Thus

$$T^{(\ell)} = K^{(\ell)} \frac{1}{1-iQK^{(\ell)}}$$

$$= \frac{1}{D}\begin{pmatrix} \alpha k^{2\ell}(1-i\gamma q^{2\ell+1})+i\beta^2(kq)^{2\ell}q & \beta(kq)^{\ell} \\ \beta(kq)^{\ell} & \gamma q^{2\ell}(1-i\alpha k^{2\ell+1})+i\beta^2(kq)^{2\ell}k \end{pmatrix},$$

where $D = (1-i\gamma q^{2\ell+1})(1-i\alpha k^{2\ell+1}) - (i\beta)^2(kq)^{2\ell+1}$. Thus the elastic amplitude is

$$T_{11}^{(\ell)} = \frac{\alpha k^{2\ell} + \dfrac{i\beta^2(kq)^{2\ell}q}{1-i\gamma q^{2\ell+1}}}{1 - ik\left[\alpha k^{2\ell} + \dfrac{i\beta^2(kq)^{2\ell}q}{1-i\gamma q^{2\ell+1}}\right]} \qquad 16.3$$

and the inelastic,

$$T_{12}^{(\ell)} = \frac{\dfrac{\beta(kq)^{\ell}}{1-i\gamma q^{2\ell+1}}}{1-ik\left[\alpha k^{2\ell} + \dfrac{i\beta^2(kq)^{2\ell}q}{1-i\gamma q^{2\ell+1}}\right]} \qquad 16.4$$

Equations 16.3 and 16.4 become more familiar if we define the "channel scattering lengths"

$$A = \alpha k^{2\ell} + \frac{i\beta^2 (kq)^{2\ell} q}{1-i\gamma q^{2\ell+1}} \qquad 16.5$$

and

$$B = \frac{\beta (kq)^{\ell}}{1-i\gamma q^{2\ell+1}} \qquad 16.6$$

Note that both A and B are complex, and both depend on k and q. We may thus write

$$T_{11}^{(\ell)} = \frac{A}{1-ikA} = \frac{1}{\frac{1}{A}-ik} \quad \text{and} \quad T_{12}^{(\ell)} = \frac{B}{1-ikA} \qquad 16.7$$

Recall that in low engery elastic scattering, we parameterize the scattering amplitude as

$$T_{11}^{(\ell)} = \frac{1}{k \cot \delta - ik} = \frac{k^{2\ell}}{k^{2\ell+1} \cot \delta - ik^{2\ell+1}} .$$

In the zero range approximation, we expect that $k^{2\ell+1}$ cot = $\frac{1}{a}$ = constant. In terms of the K-matrix, purely elastic scattering would mean that $\beta = 0$. In that case Equations 16.7 and 16.5 imply that $k \cot \delta = \frac{1}{\alpha k^{2\ell}}$ or $k^{2\ell+1} \cot \delta = \frac{1}{\alpha}$. It is precisely this idenitification with low energy elastic scattering that suggests the parameterization

$$K^{(\ell)} = Q^{\ell} \begin{pmatrix} \alpha & \beta \\ \beta & \gamma \end{pmatrix} Q^{\ell}.$$

Similarly notice that the leading term (Equation 16.4) for the inelastic amplitude goes as $\beta (kq)^{\ell}$.

From Equations 15.2 and 15.9, we may calculate the partial wave cross sections with the aid of the orthogonality of the Legendre functions. Thus

$$\sigma_{11}^{(\ell)} = 4\pi(2\ell+1)\,|T_{11}^{(\ell)}|^2 = \frac{4\pi(2\ell+1)\,|A|^2}{1+2k\,\mathrm{Im}A+k^2(A)^2}$$

$$\sigma_{12}^{(\ell)} = 4\pi\frac{q}{k}(2\ell+1)\,|T_{12}^{(\ell)}|^2 = \frac{4\pi(2\ell+1)\frac{q}{k}|B|^2}{1+2k\,\mathrm{Im}A+k^2\,|A|^2},$$

and for the total cross section,

$$\sigma_{tot}^{(\ell)} = (2\ell+1)\frac{4\pi}{k}\,\mathrm{Im}T_{11}^{(\ell)} = \frac{4\pi(2\ell+1)}{k}\;\frac{\mathrm{Im}A+k\,|A|^2}{1+2k\,\mathrm{Im}A+k^2\,|A|^2}.$$

Observing that $\mathrm{Im}A = |B|^2$, we readily verify that

$$\sigma_{tot}^{(\ell)} = \sigma_{11}^{(\ell)} + \sigma_{12}^{(\ell)}.$$

Comments:

1) Although A and B are complex functions of q and k, they are quite often parameterized as complex constants. In this case Equation 16.7 becomes a complex scattering length approximation.

2) For $\ell = 0$, $|B|^2 = \dfrac{\beta^2}{1+\gamma^2 q^2}$. Thus as $k \to 0$

$\sigma_{tot}^{(\ell)} \to \frac{4\pi}{k}\beta^2 \to \infty$. This is not a mathematical accident. Consider, for example, $K^-p \to K^-p$ and $K^-p \to \pi\Sigma$. The $\pi\Sigma$ channel is open below the K^-p threshold $(k = 0)$. Hence if the K^- comes to rest near a proton it will always interact through the $\pi\Sigma$ channel. This is true to the extent that the K^- lifetime is long compared to the time scale of the strong interaction $K^-p \to \pi\Sigma$.

3) The elastic cross section, $\sigma_{11}^{(\ell)}$, approaches a constant as $k \to 0$. Hence the $1/k$ behavior of $\sigma_{tot}^{(\ell)}$ is

due to the inelastic cross section.

4) From Equation 16.3 the elastic partial wave amplitude (Argand diagram) is

$$a_{11}^{(\ell)} = k^{1/2} T_{11}^{(\ell)} k^{1/2} \approx \alpha k^{2\ell+1} + i\ \beta^2 (kq)^{2\ell+1}$$

to leading orders in k and q. For the S-wave this is just

$$a_{11}^{(0)} \approx \alpha k + i\ \beta^2 kq.$$

In general we have $q \propto (E-E_0)^{1/2}$ where E_0 is the threshold for the second channel ($\pi\pi \to \bar{K}K$). Below this threshold the proper analytic continuation of q is $i|q|$. Thus

$$a_{11}^{(0)} \approx \begin{cases} \alpha k - \beta^2 k|q| & , E < E_0 \\ \alpha k + i\beta^2 kq & , E > E_0 \end{cases}$$

Below threshold $a_{11}^{(0)}$ is purely real and increasing since $|q| \to 0$. At threshold it suddenly becomes complex and in fact <u>must make a sharp left hand (90°) turn in the Argand plot</u>. This

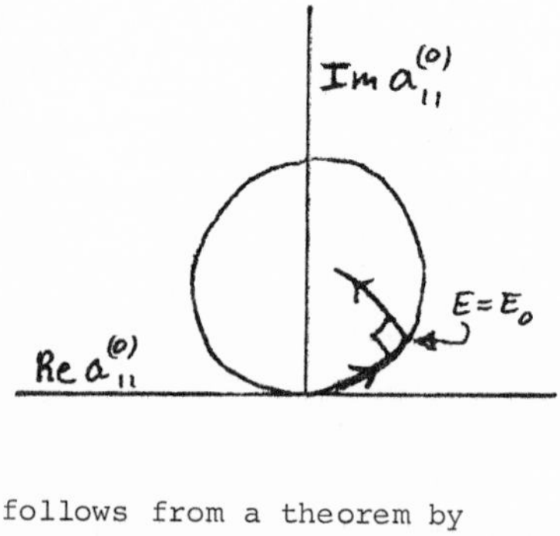

effect is quite general and follows from a theorem by Wigner, based on causality, which states that all "rapid" variations in the Argand diagram (resonances and opening thresholds!) must be in a counter-clockwise direction. Note that this result is perfectly general in the K-matrix formalism so long as $q \to i|q|$ for $E < E_0$.

In principle such effects--known as cusps--are observable in the open-channel cross section at the energy E_0 where the closed-channel opens. However, in practice experimental errors usually mask the cusp. Similar effects exist in higher partial waves but these are further suppressed by the $q^{2\ell}$ factor.

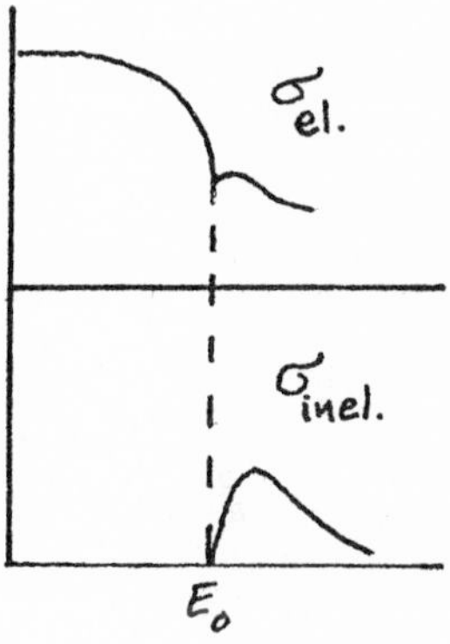

References

H. Pilkuhn, The Interaction of Hadrons, John Wiley & Sons, 1967, Chapter 3.

R.H. Dalitz, Ann. Rev. Nucl. Sci. 13, 339(1963); Rev. Mod. Phys. 33, 471(1961); especially Strange Particles and Strong Interactions, Oxford University Press, 1962.

17

THE REDUCED K-MATRIX

In this lecture we discuss the 'proper' behavior of the K-matrix as we pass through a threshold. For $E > 2m_K$ it is clear that the K-matrix for the processes $\pi\pi \to \pi\pi, \bar{K}K$ must be 2×2. However, for $E < 2m_K$ it is also clear that K is only one real number. What is the relation between these two K-matrices?

The answer to this question follows from a consideration of the analytic continuation of the T-matrix below the threshold $E = 2m_K$. Since we wish the T-matrix to have correct analytic properties, it becomes essential to understand the K-matrix behavior as we pass through a threshold.

Another reason for studying the reduced K-matrix is, perhaps, the fact that we experimentally never have access to all the open channels. Suppose we have all the data for the processes $\pi\pi \to \pi\pi, \bar{K}K$ up to an energy $E_0 < 2m_\rho (m_\rho =$ rho mass). Above $2m_\rho$ we would clearly need a 3×3 K-matrix. Below $2m_\rho$ the 2×2 K-matrix does in principle have an energy dependence which is specifically related to the strength of the closed channel ($2m_\rho$). The best we can do is to hope that at energies "sufficiently below" the two rho threshold this energy dependence can be neglected.

Extended Unitarity

Our relation for two body unitarity presently reads (cf. Equation 15.10)

$$\text{Im}T = T^{\dagger}\ Q\ T \qquad 17.1$$

In a two channel problem we have for the elastic channel

$$\text{Im}T_{11} = k|T_{11}|^2 + q|T_{12}|^2.$$

However when the second channel is closed $(q^2 < 0)$, we know that T_{11} may be written

$$T_{11} = \frac{1}{k}\, e^{i\delta} \sin \delta,$$

where δ is real. This amplitude gives

$$\text{Im}T_{11} = \frac{1}{k} \sin^2 \delta = k\left|\frac{1}{k}\, e^{i\delta} \sin \delta\right|^2 = k|T_{11}|^2$$

which implies that Equation 17.1 is incorrect for $q^2 < 0$ or $T_{12} \equiv 0$.

The correct statement of Equation 17.2 is

$$\text{Im}T = T^{\dagger}\ (Q^2) Q\ T \qquad 17.2$$

where $\theta(Q^2) = \begin{pmatrix} \theta(k^2) & 0 \\ 0 & \theta(q^2) \end{pmatrix}$. The physical origin of these step functions is that the phase space for the second channel is zero at energies below its threshold. Formally this can be seen from Equation 15.5 where the sum over intermediate states includes delta functions corresponding to energy-momentum conservation between the initial and intermediate states.

We now rewrite Equation 17.2 in order to see whether it says anything about the K-matrix.

$$\frac{T-T^{\dagger}}{2i} = T^{\dagger}\theta(Q^2)Q\,T,$$

$$\frac{(T^{\dagger})^{-1}-T^{-1}}{2i} = \theta(Q^2)Q \qquad 17.3$$

or

$$\mathrm{Im}(T^{-1}) = -\theta(Q^2)Q$$

From $T = K(1-iQK)^{-1}$ we also have

$$K^{-1} = T^{-1} + iQ \qquad 17.4$$

or

$$\mathrm{Im}(K^{-1}) = \mathrm{Im}(T^{-1}) + \mathrm{Re}Q$$

Below the second channel threshold, q becomes pure imaginary $(q^2 < 0)$; that is,

$$K^{-1} = T^{-1} + \begin{pmatrix} ik & 0 \\ 0 & -|q| \end{pmatrix}$$

or $$\mathrm{Im}(K^{-1}) = \mathrm{Im}(T^{-1}) + \begin{pmatrix} k & 0 \\ 0 & 0 \end{pmatrix}.$$

Thus from Equation 17.3 we conclude that $\mathrm{Im}(K^{-1}) = 0$ both above and below the threshold--the appropriate K-matrix is a real function of energy on the real axis.

Comments:

1) Below a threshold we take $q \to +i|q|$. This is because we are studying the T and K matrices for physical scattering on the real axis of the first Riemann sheet.

2) For an N dimensional problem, there are 2^N sheets associated with T. This is simply because T is a function of N momenta each of which introduces a cut or two sheets.

The Reduced K-Matrix

We begin this section by observing that any reasonably parameterized K-matrix becomes a complex function of energy below a threshold. Thus our canonical example,

$$K^{(\ell)} = Q^{\ell} \begin{pmatrix} \alpha & \beta \\ \beta & \gamma \end{pmatrix} Q^{\ell},$$

becomes complex since $q^{\ell} \to (i|q|)^{\ell}$ when $q^2 < 0$. The difficulty is that when we say the K-matrix is real we actually mean the reduced K-matrix is real.

To find the reduced K-matrix we first write the full K matrix as

$$K = \begin{pmatrix} K_{oo} & K_{oc} \\ K_{co} & K_{cc} \end{pmatrix},$$

where the subscripts o,c denote open, closed channels. For the full T-matrix we have, using

$$Q = \begin{pmatrix} Q_o & o \\ o & Q_c \end{pmatrix},$$

$$\begin{pmatrix} T_{oo} & T_{oc} \\ T_{co} & T_{cc} \end{pmatrix} = \begin{pmatrix} K_{oo} & K_{oc} \\ K_{co} & K_{cc} \end{pmatrix} \begin{pmatrix} 1-iQ_oK_{oo} & -iQ_oK_{oc} \\ -iQ_cK_{co} & 1-iQ_cK_{cc} \end{pmatrix}^{-1}.$$

We wish to write T_{oo} in terms of K_{oo}, K_{oc}, K_{cc}. Since K_{oo} is in general a matrix, it is easier to use

$$T(1-iQK) = K$$

to compute T_{oo}. This equation gives the two matrix equations

$$T_{oo}(1-iQ_oK_{oo}) - iT_{oc}Q_cK_{co} = K_{oo}$$

$$-iT_{oo}Q_oK_{oc} + T_{oc}(1-iQ_cK_{cc}) = K_{oc} .$$

T_{oc} may readily be eliminated from these equations (Q_cK_{cc} is a number, not a matrix) to give

$$T_{oo} = K_r(1-iQ_oK_r)^{-1} \qquad 17.6$$

where

$$K_r = K_{oo} + \frac{iQ_cK_{oc}K_{co}}{1-iQ_cK_{cc}} \qquad 17.7$$

Equation 17.7 is known as the reduced K-matrix. When $Q_c^2 > 0$ (above "threshold"), K_r is in general complex, while the full K-matrix is of course real. Below threshold, where $Q_c^2 < 0$, the reduced K-matrix becomes

$$K_r = K_{oo} - \frac{|Q_c|K_{oc}K_{co}}{1+|Q_c|K_{cc}} .$$

Since K_{oo}, K_{oc}, K_{cc} are real, it is clear that K_r is real below threshold.

Comments:

1) Below threshold extended unitarity implies

$$\mathrm{Im}T_{oo} = T^+_{oo}Q_oT_{oo} \quad \text{or} \quad \mathrm{Im}(T^{-1}_{oo}) = -Q_o .$$

Equation 17.6 implies $\mathrm{Im}(K_r^{-1}) = \mathrm{Im}(T^{-1}_{oo}) + Q_o$; but this is consistent since K_r and K_r^{-1} are real below threshold. We thus conclude that the T matrix given by K are consistent both above and below threshold with extended unitarity.

2) Notice that a pole can occur in K_r provided

$1 + |Q_c|K_{cc} = 0$ or $K_{cc} = -1/|Q_c|$.

This simple fact has led to a great deal of theoretical discussion concerning the dynamical origin of resonances near opening thresholds. We shall see later that such a pole leads naturally to resonance behaviour in T_{oo}. Clearly the location of the pole ($\approx$ mass of resonance) depends on the "strength" of the closed channel (K_{cc}). From this point of view it is not surprising that many of the better known inelastic N^* and Y^* resonances occur at energies corresponding to quasi-two-body thresholds.

References

H. Pilkuhn, The Interaction of Hadrons, John Wiley & Sons, 1967. Chapter 8, but considerably expanded.

18

THE K-MATRIX AND RESONANCES

We have seen that a pole can arise in the reduced K matrix when $1 + |Q_c| K_{cc} = 0$. We shall study in this lecture how such a pole can lead to a Breit-Wigner behavior in T_{oo}. From an "effective-range expansion" of K^{-1} (analogous to $k \cot \delta$), we shall also learn that poles in the full K matrix ($\det K^{-1} = 0$) also lead to a resonant interpretation in T.

After discussing some additional features of the K matrix (eigenphases, "factorization" near a pole, etc.), we shall indicate some limitations of the K-matrix parameterizations. In particular we present a K-matrix which has no poles, whose reduced K-matrix has no poles, but whose corresponding T matrix does have a pole! Thus while the presence of poles in K (or K_r) implies poles in T, the converse does not hold in general.

Poles in K_R: Resonant Interpretation

For simplicity of discussion we consider a two-channel problem with the constant matrix

$$K = \begin{pmatrix} K_{oo} & K_{oc} \\ K_{oc} & K_{cc} \end{pmatrix}$$

We take k to be the open channel momentum and q ($= i|q|$ below threshold) to be the "closed" channel momentum; we assume the particles in the two channels to be of equal mass ($\pi\pi \rightarrow \pi\pi$, $\pi\pi \rightarrow \bar{K}K$, $m_\pi \neq m_K$!).

From Equations 17.6 and 17.7, we have below threshold

$$T_{oo} = \frac{K_r}{1-ikK_r} \qquad 18.1$$

where

$$K_r = K_{oo} - \frac{|q|K_{oc}^2}{1+|q|K_{cc}} \qquad 18.2$$

We assume that

$$1+|q|K_{cc} = 0 \qquad 18.3$$

at some energy $E_o < 2m_K$. This assumption implies $K_{cc} < 0$. Expanding $1+|q|K_{cc}$ about $E = E_o$, we have

$$1+|q|K_{cc} \approx (E-E_o)\frac{d|q|}{dE}\bigg|_{E=E_o} K_{cc} = (E_o-E)\frac{E_o}{4|q_o|} K_{cc}, \qquad 18.4$$

since $|q| = (4m_k^2-E^2)^{1/2}/2$. Inserting Equation 18.4 into Equation 18.2, we see explicitly that K_r has a pole at $E = E_o$,

$$K_r = K_{oo} - \frac{|q_o|K_{oc}^2}{(E_o-E)\frac{E_o}{4|q_o|}K_{cc}} .$$

To proceed we assume that K_{oo} is small compared to the residue of the pole term. This assumption does not invalidate the presence of a (complex) pole in T_{oo}, but it does make a Breit-Wigner identification more transparent.

We therefore have

$$T_{oo} = \frac{1}{k} \frac{-\dfrac{k|q_o|K_{oc}^2}{(E_o-E)\frac{E_o}{4|q_o|}K_{cc}}}{1 + \dfrac{ik|q_o|K_{cc}^2}{(E_o-E)\frac{E_o}{4|q_o|}K_{cc}}} = \frac{1}{k}\frac{\Gamma/2}{E_o-E-i\Gamma/2}\,, \qquad 18.5$$

where $\Gamma/2 = \frac{-4k|q_o|^2K_{oc}^2}{E_oK_{cc}}$. Recall that $K_{cc} < 0$ so that $\Gamma > 0$.

We next look at the closed-channel matrix element T_{cc},

$$T_{cc} = \frac{K_{cc} + \dfrac{iK_{oc}^2k}{1-ikK_{oo}}}{1 + |q|\left(K_{cc} + \dfrac{iK_{oc}k}{1-ikK_{oo}}\right)} \qquad 18.6$$

$$\approx \frac{K_{cc} + iK_{oc}^2k}{(E_o-E)\frac{E_o}{4|q_o|}K_{cc} + i|q_o|kK_{oc}^2}$$

where we again neglect K_{oo}. Thus,

$$T_{cc} \propto \frac{1}{E_o - E - i\Gamma/2} \qquad 18.7$$

so that T_{cc} also has a pole at $E \approx E_o - i\Gamma/2$.

Comments:

1) The best known example of a pole in the reduced K-matrix (Equations 18.5 and 18.6) is the $Y_o^*(1405)$. In this case T_{oo} corresponds to $\pi\Sigma \to \pi\Sigma$ scattering (not directly measurable--no Σ targets!) and T_{cc}, to $\bar{K}N \to \bar{K}N$. E_o is, of course, 1405 which is some 15 MeV below

threshold. If one fits $\bar{K}N \to \pi\Sigma$ and $\bar{K}N \to \bar{K}N$ data just above threshold with the K-matrix, then a typical resonant structure is observed in ImT_{cc} (and ReT_{cc}) when computed

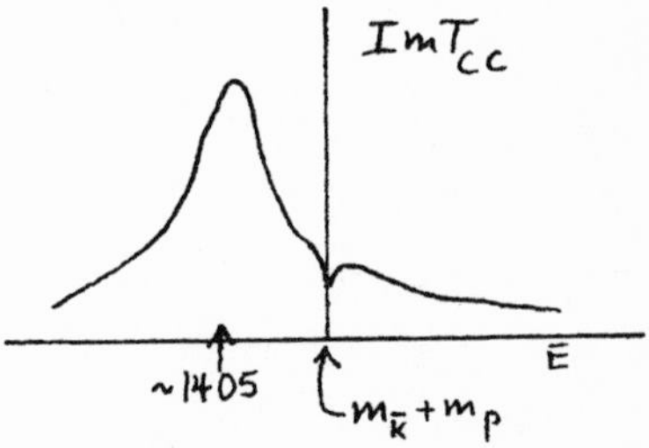

below the elastic threshold. In production experiments this resonance is only seen in $\pi\Sigma$ mass plots $(M_{\pi\Sigma} < 1420)$, consistent with Equation 18.5 which implies it should be essentially an elastic resonance in the $\pi\Sigma$ channel $(\Gamma_{\pi\Sigma} \cong \Gamma_{total})$. Such a pole is sometimes referred to as a "virtual bound state" since it cannot decay into the closed channel $(\bar{K}N)$.

2) Suppose we were only discussing a one channel problem for T_{cc}. Then, of course, $K_{oo} = 0$ and $K_{oc} = 0$ (i.e., $\Gamma = 0!$). In this case we have below threshold

$$T_{cc} = \frac{K_{cc}}{1+|q|K_{cc}} = \frac{1}{\frac{1}{K_{cc}} + |q|}$$

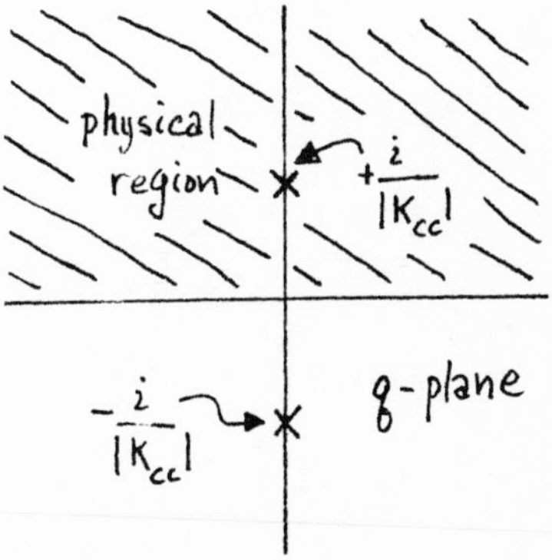

It is clear that T_{cc} can have poles at

a) $i|q| = \frac{i}{|K_{cc}|}$, $K_{cc} < 0$

b) $i|q| = -\frac{i}{|K_{cc}|}$, $K_{cc} > 0$.

The classic example of these possibilities is np scattering in the S-wave. It turns out that case a) corresponds to the ^3S state, and the pole corresponds to the deuteron which is stable simply because $K_{oc} \equiv 0$ ($\Gamma=0$!). Case b) corresponds the the ^1S or "virtual" state. The only reason (not very fundamental!) we never refer to the ^1S state as a stable particle is that the pole is in the unstable region of the q-plane, whereas the deuteron pole is on the physical sheet, connected to the observable world. If there were only a T_{oo} ... !

Parameterizations of K

The examples of K so far studied have involved the parameterization

$$K^{(\ell)} = Q^{\ell} \begin{pmatrix} \alpha & \beta \\ \beta & \gamma \end{pmatrix} Q^{\ell}$$

for the ℓth partial wave with α,β,γ constant. We have seen that this parameterization is a natural analogue to the scattering length approximation, $k^{2\ell+1} \cot \delta_\ell = 1/a_\ell$. We now give α,β,γ an energy dependence in a fairly general manner, through the so-called "M-matrix" parameterization.

Take the following equation as a definition of M:

$$K^{-1} = Q^{-\ell} M Q^{-\ell} \qquad 18.8$$

where as usual $Q^{\ell} = \begin{pmatrix} k_1^{\ell} & 0 & \cdots \\ 0 & k_2^{\ell} & \\ \vdots & & \ddots \end{pmatrix}$ for the ℓth partial wave. Since

$$T = K \frac{1}{1-iQK},$$

we have

$$T(1-iQ^{\ell+1}M^{-1}Q^{\ell}) = Q^{\ell}M^{-1}Q^{\ell}$$

$$TQ^{-\ell}Q^{\ell}(Q^{-\ell}-iQ^{\ell+1}M^{-1}) = Q^{\ell}M^{-1}$$

$$TQ^{-\ell}(M-iQ^{2\ell+1}) = Q^{\ell}$$

or
$$T = Q^{\ell}\,\frac{1}{M-iQ^{2\ell+1}}\,Q^{\ell}. \qquad 18.9$$

The parallel between this matrix equation and the one channel expression is clear:

$$T_{\text{one channel}} = \frac{1}{k}\,e^{i\delta}\sin\delta$$

$$= \frac{1}{k\cot\delta - ik}$$

$$= \frac{k^{2\ell}}{k^{2\ell+1}\cot\delta - ik^{2\ell+1}} \qquad 18.10$$

Schematically,

$$\begin{array}{ccl} k^{2\ell+1}\cot\delta & \Longleftrightarrow & M \\ \updownarrow & & \updownarrow \\ \frac{1}{a} + \frac{1}{2}rk^2 & \Longleftrightarrow & M^{(0)} + R(E-E_o) \;. \end{array}$$

The parallel between Equations 18.9 and 18.10 suggests we write

$$M = \begin{pmatrix} M_{11}^{(0)} + R_{11}(E-E_o) & M_{12}^{(0)} + R_{12}(E-E_o) \\ \swarrow\!\!\nearrow & M_{22}^{(0)} + R_{22}(E-E_o) \end{pmatrix}, \qquad 18.11$$

where $M^{(0)}$ and R are symmetric, real, constant matrices. In effect we expand the M-matrix about some energy E_o just as $k^{2\ell+1}\cot\delta$ is expanded about zero momentum.

Comments:

1) One can prove in (Schrödinger) potential theory that the derivatives $\frac{dM}{dE}$ exist at E_o. To this extent Equation 18.11 is theoretically justified.

2) We often see M written as

$$M = M^{(0)} + R \begin{pmatrix} k_1^2 - k_{1(0)}^2 & 0 \\ 0 & k_2^2 - k_{2(0)}^2 \end{pmatrix},$$

Where (0) denotes the (common) energy E at which $M = M^{(0)}$. Since in potential theory (and approximately for relativistic kinematics) $E \propto k^2$, there is little difference between this expansion and that of Equation 18.11.

If it happens that the determinant of M has two real zeroes in E--note that det M is just a quadratic equation in E--then we may easily show that the T-matrix will have Breit-Wigner-like poles. By assumption, then,

$$\det M = M_{11}M_{22}-M_{12}^2 = C(E-E_1)(E-E_2), \qquad 18.12$$

where $C = R_{11}R_{22}-R_{12}^2$. Notice that the K matrix, which is proportional to det M, will in general have poles at E_1, E_2.

To proceed we let $E_1 = E_R(\neq E_2)$ and set $C' = C(E_R-E_2)$. Thus near $E = E_R$ we have

$$\left.\begin{aligned} &\det M \cong C'(E-E_R), \\ &M_{11}(E_R)M_{22}(E_R) = M_{12}^2(E_R). \end{aligned}\right\} \qquad 18.13$$

From Equations 18.8 and 18.13 we have

$$K = Q^{\ell} \frac{1}{\det M} \begin{pmatrix} M_{22} & -M_{12} \\ -M_{12} & M_{11} \end{pmatrix} Q^{\ell}$$

$$\approx Q^{\ell} \frac{1}{C'(E-E_R)} \begin{pmatrix} M_{22}(E_R) & \pm\left[M_{11}(E_R)M_{22}(E_R)\right]^{1/2} \\ \swarrow & M_{11}(E_R) \end{pmatrix} Q^{\ell} \qquad 18.14$$

Finally, assuming $C' < 0$ we have for the T-matrix

$$T_{ij} = \left(K \frac{1}{1-iQK}\right)_{ij} = \frac{1}{(k_i k_j)^{1/2}} \frac{\frac{1}{2}(\Gamma_i \Gamma_j)^{1/2}}{E_R - E - i\Gamma_{tot}/2}, \qquad 18.15$$

where

$$\left.\begin{aligned} \Gamma_1 &= 2k_1^{2\ell+1} \left(\frac{M_{22}(E_R)}{|C'|}\right)^2 \\ \Gamma_2 &= 2k_2^{2\ell+1} \left(\frac{M_{11}(E_R)}{|C'|}\right)^2 \\ \Gamma_{tot} &= \Gamma_1 + \Gamma_2 \end{aligned}\right\} \qquad 18.16$$

Comments:

1) The elements of the K-matrix, Equation 18.14, have a mutual pole at $E = E_R$ where $\det M = 0$. Furthermore, the residues of these poles factorize in the sense that $M_{11}(E_R)M_{22}(E_R) = M_{12}^2(E_R)$.

2) The "partial widths," Equation 18.16, have the correct threshold dependence, $k_i^{2\ell+1}$.

3) The M's and C' are, of course, functions of energy; so then is $\Gamma_i/k_i^{2\ell+1}$. Note the extreme situation when $E = E_2$ so that $C' = 0$! Note that $C' < 0$ is assumed to insure that the amplitudes have a counter-clockwise motion in the Argand plot. This sign choice is related to a similar situation in the usual effective range expansion; namely, if the relative sign between a and r is properly chosen, $k^{2\ell+1} \cot \delta = \frac{1}{a} + \frac{1}{2} rk^2$ can pass through zero.

4) We have seen that a pole in the reduced K-matrix leads to a resonance. In some sense this resonance has its dynamic origin from the closed channel which "causes" the pole in $K_r(\alpha \frac{1}{1+|q_c|K_{cc}})$. What, however, is the dynamic origin of a resonance related to a pole in the full K-matrix? Recall that what is called the "full K-matrix" is actually the reduced K-matrix of some higher channel problem. Thus when we use the M-matrix parameterization, we are actually making an approximation to the correct energy dependence of a reduced K-matrix. Hence one dynamical interpretation of a pole in the "full K-matrix" is that it is actually due to a channel "far" above the pole location in energy.

5) Finally we mention that Wigner and Eisenbud showed long ago in Schrödinger potential theory that the K-matrix can be in general written as

$$(Q^{-\ell} K Q_j^{-\ell})_{ij} = \sum_R \frac{\gamma_{iR}\gamma_{jR}}{E_R - E} ,$$

where the γ's are constants. Notice that for one term in the sum, this corresponds exactly to Equation 18.14. Whether this relation holds in a relativistic theory is an open question.

The Eigenphase Representation

Since K is real and symmetric, so is

$$K' = Q^{1/2}\, K\, Q^{1/2} \left(= Q^{\ell+1/2} M^{-1} Q^{\ell+1/2}\right) \qquad 18.17$$

above all thresholds for the q_i in $Q = \begin{pmatrix} q_1 & 0 & \cdots \\ 0 & q_2 & \\ \vdots & & \ddots \end{pmatrix}$.

Below a threshold a similar statement holds for K_R; in general, when we discuss the eigenphase representation of K we actually mean the appropriate K_R.

The effect of introducing K' is to "absorb" all the Q matrices in expression for T and S. Thus

$$T = K \frac{1}{1-iQK}$$

$$Q^{1/2} T (Q^{1/2} Q^{-1/2}) (1-iQK) = Q^{1/2} K$$

$$Q^{1/2} T\, Q^{1/2} (1-iQ^{1/2} K Q^{1/2}) = Q^{1/2} K\, Q^{1/2},$$

or

$$\left.\begin{aligned} T' &= K' \frac{1}{1-iK'} \\ \text{where} \quad T' &= Q^{1/2} T\, Q^{1/2} \end{aligned}\right\} \qquad 18.18$$

Clearly, $S = 1 + 2i\,T'$. Note that if we are below a threshold then $T' \to T'_R = K'_R \frac{1}{1-i\,K'_R}$. This is intuitively obvious since $S_{cc} = 1$ for the closed channel.

Since K' (or K'_R) is real and symmetric, there exists a real, orhogonal matrix $C(C^TC = 1)$ such that

$$C^T K' C = K'_D \tag{18.19}$$

where K'_D is diagonal. C also diagonalizes T':

$$C^T \; T'(CC^T)(1-iK') = C^T K'$$

or
$$C^T \; T'C = K'_D(1-iK'_D)^{-1},$$

the right-hand side is diagonal since the product of two diagonal matrices is diagonal.

Since $S = 1 + 2iT'$, we have

$$S_D = C^T \; S \; C \tag{18.20}$$

where S_D is diagonal. From the unitarity of S we also may write

$$(S_D)_{ij} = \delta_{ij} \; e^{2i\delta_i}$$

or
$$S_D = \begin{pmatrix} e^{2i\delta_1} & 0 & 0 & \cdots \\ 0 & e^{2i\delta_2} & 0 & \cdots \\ 0 & 0 & e^{2i\delta_3} & \\ \vdots & \vdots & & \ddots \end{pmatrix}, \tag{18.21}$$

where the δ_i--the so-called eigenphases--are all real numbers. Finally, from

$$S_D = 1 + 2iK'_D \frac{1}{1-iK'_D} = (1+iK'_D) \frac{1}{(1-iK'_D)}$$

we have

$$K'_D = -i(S_D+1)^{-1}(S_D-1).$$

With the explicit representation, Equation 18.21, we have

$$(K_D')_{ij} = -i\delta_{ij}\,\frac{e^{2i\delta_i}-1}{e^{2i\delta_i}+1} = \delta_{ij}\tan\delta_i ; \qquad 18.22$$

that is, the eigenvalues of K' (elements of K_D') are just the tangents of the eigenphases of S_D, the δ_i.

We have introduced the eigenphase representation primarily for the sake of completeness. Although it was once believed that a resonance would manifest itself by an eigenphase passing through 90°, it is now known that this is not in general the case. The ultimate criterion for resonance remains the requirement that the T-matrix have a pole in the complex E-plane. We shall persue the discussion a bit further, however, to show if some δ_i passes through 90° (K has a pole) then the T-matrix does have a nearby pole. Thus the criterion that δ_R pass through 90° remains valid, but requires the other eigenphases to be slowly varying and of such a magnitude as to nowhere in E cross the "resonating phase".

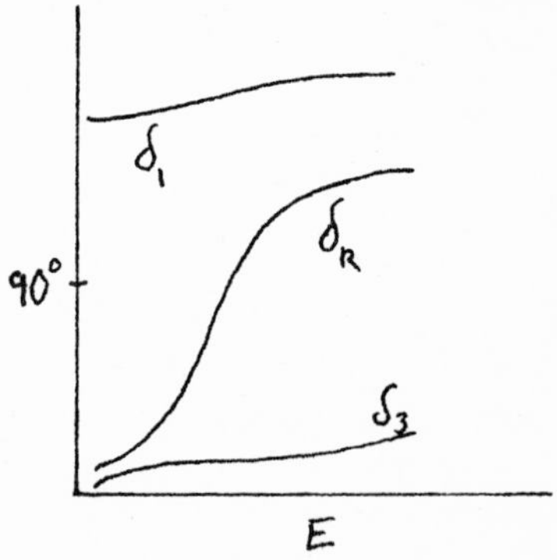

The matrix C relates the physical, open channels $|i\rangle$, $|f\rangle$ to the eigen channels $|\alpha\rangle$, $|\beta\rangle$:

$$|\alpha\rangle = \sum_i C_{\alpha i}|i\rangle . \qquad 18.23$$

Once we find the δ_i, we may construct C and return to the physical channel T-matrix.

$$T'_{if} = (C\, T'_D\, C^T)_{if}$$

$$= \sum_{\alpha\beta} C_{i\alpha} (T'_D)_{\alpha\beta}\, C_{f\beta}$$

$$= \sum_{\alpha\beta} C_{i\alpha}\, \delta_{\alpha\beta}\, e^{i\delta_\alpha} \sin\delta_\alpha\, C_{f\beta}$$

$$= \sum_{\alpha} C_{i\alpha}\, C_{f\alpha}\, e^{i\delta_\alpha} \sin\delta_\alpha \qquad 18.24$$

Suppose one of the eigenphase, δ_R, passes through 90°, while the others are negligible. Then

$$T'_{if} = \frac{C_{iR}C_{fR}}{\cot\delta_R - i},$$

where $\cot\delta_R \approx \frac{E_R - E}{\Gamma/2}$, $\Gamma/2 \approx -\left.\frac{d\cot\delta_R}{dE}\right|_{E=E_R}$.

This equation may be written

$$T'_{if} = \frac{\sqrt{\Gamma_i\Gamma_f}/2}{E_R - E - i\Gamma/2}, \qquad 18.25$$

where $\Gamma_i = C_{iR}^2\Gamma$. Since C is orthogonal, $\sum_i C_{iR}^2 = 1$ or $\sum_i \Gamma_i = \Gamma$.

Equation 18.25 may be generalized somewhat through the observation that while $\cot\delta = (E_R - E)/(\Gamma/2)$ has a zero, the expression

$$\cot\delta = \frac{(E^* - E)}{\gamma/2} + \cot\delta_B$$

could also have a zero $E_R (\neq E^*)$. Thus the presence of a (constant?) background $\cot \delta_B$ would shift the zero in $\cot \delta$ from the location of the resonance.

Suppose our resonant eigenvalue is

$$\tan \delta = \frac{\gamma/2}{E^* - E} + \tan \delta_B ,$$

where γ and δ_B are constant. If in reality

$$\delta = \delta_R + \delta_B ,$$

where $\tan \delta_R = \frac{\Gamma/2}{E_R - E}$, then one readily shows that

$$\left. \begin{aligned} E_R &= E^* + \frac{\gamma}{4} \sin 2\delta_B \\ \frac{\Gamma}{2} &= \frac{\gamma}{2} \cos^2 \delta_B \end{aligned} \right\} \qquad 18.26$$

Hence the actual resonant mass (E_R) and width (Γ) can be quite different from the apparent values (E^*, γ) even in the eigenphase representation. In practice the situation is much worse since γ and δ_B are generally ("strong") functions of the energy.

Limitations of the K-Matrix Formalism

The K-matrix formalism we have discussed is strictly applicable only to two-body final states. When three (or more) particles are present the matrix equation $T(1-iQK) = K$ becomes an integral equation, so that the entire problem is no longer algebraic. In practice one either ignores the three body final states or attempts to approximate their effect by introducing an additional "quasi-two-body" channel.

Thus we might hope that a $\pi p \to \pi\Delta(1236)$ would be a good approximation to $\pi p \to \pi\pi p$. In this case one actually does observe Δ in production experiments near the $\pi\pi p$ threshold. In other cases, such as $\pi\pi \to \pi\pi\pi\pi$ ($E_{4\pi} < E_{K\bar{K}}$!), the quasiparticles or thresholds may have little to do with any produced resonances (there are no 2π resonances with mass $\approx$ 280 MeV!). In the end the best we can do is to hope the non-two-body channels are not too important; unfortunately, Nature need not be so generous.

We have noted that the eigenphase representation is not generally useful. A perhaps unfortunate side effect of the M-matrix parameterization (especially true in a two channel problem), $M = M_D + R(E-E_D)$, is that the eigenphases tend to be distinct in that one will pass through 90° and the other will be relatively constant. A more general possibility is shown in the sketch. Here all three eigenphases are rapidly varying in the region of E_0. Note that the eigenphase behave with E in such a manner as not to cross one another. This effect is known as Wigner's "no crossing" theorem

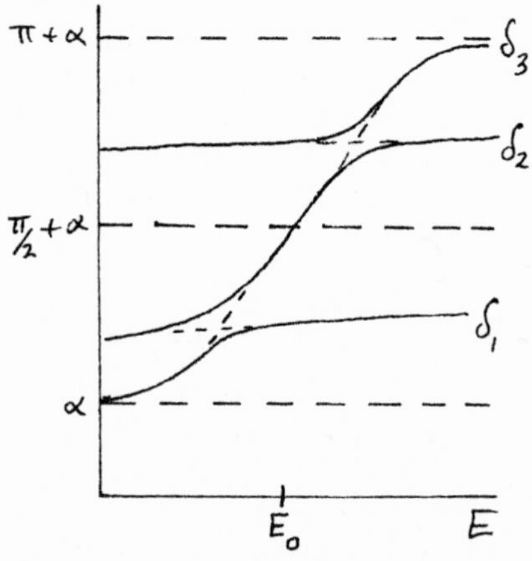

for eigenphases. A rather general proof of this theorem has been given by Weidenmüller (Phys. Letts. 24B, 441(1967).

We might still hope that at least one eigenphase passes through 90° near the resonance location. This hope will be quickly dispelled when we consider an example where T has a pole but the corresponding K-matrix does not. Since the tangents of the eigenphases are proportional to the elements of K' (they are just the eigenvalues of K'), it is clear that no δ_i will be 90° if K (or K') has no poles.

Consider the T-matrix

$$T = \begin{pmatrix} \frac{1}{k_1}\left[\frac{e^{2i\delta}-k}{2i} + \frac{e^{2i\delta}m\Gamma_1}{m^2-E^2-im\Gamma}\right] & \frac{1}{(k_1k_2)^{1/2}}\frac{e^{i\delta}m\sqrt{\Gamma_1\Gamma_2}}{m^2-E^2-im\Gamma} \\ \swarrow & \frac{1}{k_2}\frac{m\Gamma_2}{m^2-E^2-im\Gamma} \end{pmatrix}, \qquad 18.27$$

where $\Gamma_1 = g_1k_1$, $\Gamma_2 = g_2k_2$, $\Gamma = \Gamma_1 + \Gamma_2$, m is the "resonance" mass, and δ is a "background" phase.

Using $S = I + 2iQ^{1/2}\,T\,Q^{1/2}$, $Q = \begin{pmatrix} k_1 & 0 \\ 0 & k_2 \end{pmatrix}$, one easily verifies that $S^+S = SS^+ = I$.

We next compute the corresponding K matrix in the case where $\delta = 90^{\circ}$. The S matrix becomes

$$S = \begin{pmatrix} -1 - \frac{2im\Gamma_1}{M^2-E^2-im\Gamma} & \frac{-2m\sqrt{\Gamma_1\Gamma_2}}{M^2-E^2-im\Gamma} \\ \swarrow & 1 + \frac{2im\Gamma_2}{m^2-E^2-im\Gamma} \end{pmatrix}.$$

S will have poles <u>near</u> $E^2 \approx m^2 - im\Gamma$ (or $E \approx m - i\Gamma/2$). Since $T = K(I - iQK)^{-1}$, we have

$$iQ^{1/2} K Q^{1/2} = (S+I)^{-1}(S-I). \tag{18.28}$$

After some matrix algebra we may explicitly exhibit K,

$$K = \begin{pmatrix} \dfrac{E^2-m^2}{g_1 m k_1^2} & -\dfrac{1}{k_1}\sqrt{\dfrac{g_2}{g_1}} \\ -\dfrac{1}{k_1}\sqrt{\dfrac{g_2}{g_1}} & 0 \end{pmatrix} \tag{18.29}$$

The g_i are essentially the squares of (constant) coupling constants. Notice that K does not involve k_2. Also K is not singular in k_1, so long as we use Equation 18.27 far from $k_1 = 0$. Thus if the resonant energy m does not correspond to $k_1 = 0$, we see that K has no poles in the region where $E \approx m$. In addition we recall that

$$K_R = K_{11} + \frac{ik_2 K_{12}^2}{1-ik_2 K_{22}} .$$

Since $K_{22} = 0$ it is clear that the reduced K matrix has no poles about $E = m$.

At $E = m$, $K' = \begin{pmatrix} 0 & -\sqrt{\Gamma_2/\Gamma_1} \\ -\sqrt{\Gamma_2/\Gamma_1} & 0 \end{pmatrix}$; we thus have

$$\tan \delta_{\pm} = \pm \left(\frac{g_2 k_2}{g_1 k_1}\right)^{1/2} \text{(at } E = m), \tag{18.30}$$

where $\delta_\pm$ are the eigenphases. The eigenphases are shown in the sketch. Note that they are in general not 90° at E_0.

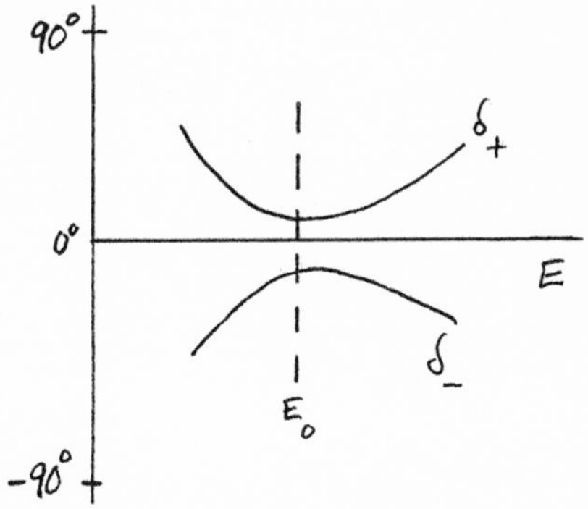

The ultimate conclusion to be drawn from this example is that one must search in the complex E plane for poles in T; poles in $K(K_R)$ generally imply (nearby) poles in T, the converse need not be so. In addition the pole in K may be quite far from Re E corresponding to the complex pole in T. Just how far depends on how large the effective "background" is.

References

The K-Matrix and Resonances:

Dalitz, see references for Lectures 15, 16.

M.H. Ross and G.L. Shaw, Am. Phys. 13,147(1961)--on the M-matrix.

S.M. Flatté, et al., Group A, LRL, preprint on S-wave scattering (to be published). Introduces the T-matrix used here as an example of a K-matrix without poles.

C.J. Goebel and K.W. McVoy, P.R. 164, 1932(1967). Eigenvalues of S from viewpoint of poles in S. Stresses that eigenphase representation is generally complicated.

H.A. Weidenmüller, Phys. Letts. 24B, 441(1967). Proof of no-crossing theorem for eigenphases.

R.H. Dalitz and R.G. Moorhouse, Proc. Roy. Soc. Lond. A., 318, 279(1970). "What is Resonance?" An elegant summary of the various manifestations of and criteria for resonances.

19

THE PION-NUCLEON INTERACTION. BARYON RESONANCES WITH ZERO STRANGENESS

General Features of π-N Cross Sections

Isospin decomposition of π-N scattering: The physical π-N states may be written in terms of the isospin states $I = 3/2$ and $I = 1/2$

$$\begin{aligned} |\pi^+p\rangle &= |3/2\rangle \\ |\pi^-p\rangle &= \sqrt{1/3}\,|3/2\rangle - \sqrt{2/3}\,|1/2\rangle \\ |\pi^on\rangle &= \sqrt{2/3}\,|3/2\rangle + \sqrt{1/3}\,|1/2\rangle \end{aligned} \qquad 19.1$$

Accordingly, the scattering amplitudes are:

$$\begin{aligned} f_+ &= \langle \pi^+p|\pi^+p\rangle = \langle 3/2|3/2\rangle = f_{3/2} \\ f_- &= \langle \pi^-p|\pi^-p\rangle = 1/3\ f_{3/2} + 2/3\ f_{1/2} \\ f_o &= \langle \pi^on|\pi^-p\rangle = \sqrt{2/3}\ f_{3/2} - \sqrt{2/3}\ f_{1/2} \end{aligned} \qquad 19.2$$

where $f_{3/2}$ and $f_{1/2}$ are the scattering amplitudes for the isospin states $I = 3/2$ and $I = 1/2$ respectively.

From 19.2 one derives for the elastic cross sections:

$$\begin{aligned} \frac{d\sigma}{d\Omega}(\pi^+p \to \pi^+p) &= |f_+|^2 = |f_{3/2}|^2 \\ \frac{d\sigma}{d\Omega}(\pi^-p \to \pi^-p) &= |f_-|^2 = 1/9|f_{3/2} + 2\ f_{1/2}|^2 \\ \frac{d\sigma}{d\Omega}(\pi^-p \to \pi^on) &= |f_o|^2 = 2/9|f_{3/2} - f_{1/2}|^2 = 1/2|f_+ - f_-|^2 \end{aligned} \qquad 19.3$$

For the total cross sections we have, from the optical theorem:

$$\sigma_T^+ = 4\pi ƛ \; \mathrm{Im} f_3(0) = \sigma_T(3/2) \qquad 19.4$$

$$\sigma_T^- = 4\pi ƛ \; \mathrm{Im}|1/3 \; f_3(0) + 2/3 \; f_1(0)| = 1/3\sigma_T(3/2) + 2/3 \; \sigma_T(1/2)$$

where f_3 stands for $f_{3/2}$ and f_1 for $f_{1/2}$. Clearly each isospin amplitude f_{2I} has the form seen previously:

$$f = g(\theta) + ih(\theta)\vec{\sigma}\cdot\hat{n} \qquad 19.5$$

From 19.4 we have

$$\sigma_T(1/2) = 1/2|3\sigma_T(\pi^- p) - \sigma_T(\pi^\dagger p)| \qquad 19.6$$

In terms of the partial wave expansion seen at page 114, and with the nomenclature for the partial waves $T_{I,\ell\pm} \equiv \ell_{2I,2J}$, the partial waves for π-N scattering are:

for $I = 1/2$, S_{11} P_{11} P_{13} D_{13} D_{15} F_{15} F_{17} G_{17} G_{19} ...

for $I = 3/2$, S_{31} P_{31} P_{33} D_{33} D_{35} F_{35} F_{37} G_{37} G_{39} ...

The behavior of the π^+p and π^-p cross sections is shown in the accompanying plots. Shown also is a plot of the πN total cross sections in the two isospin states. To be noticed is that all baryon resonances with zero strangeness have been discovered in formation experiments. In addition to the total cross sections, differential cross sections and polarization have been measured with great accuracy, over a large range of momenta. The great majority of the data on π-N scattering originates in counter experiments. As we have seen, bumps in σ_T are a good indicator of the presence of resonances. As discovered by partial wave analyses, however,

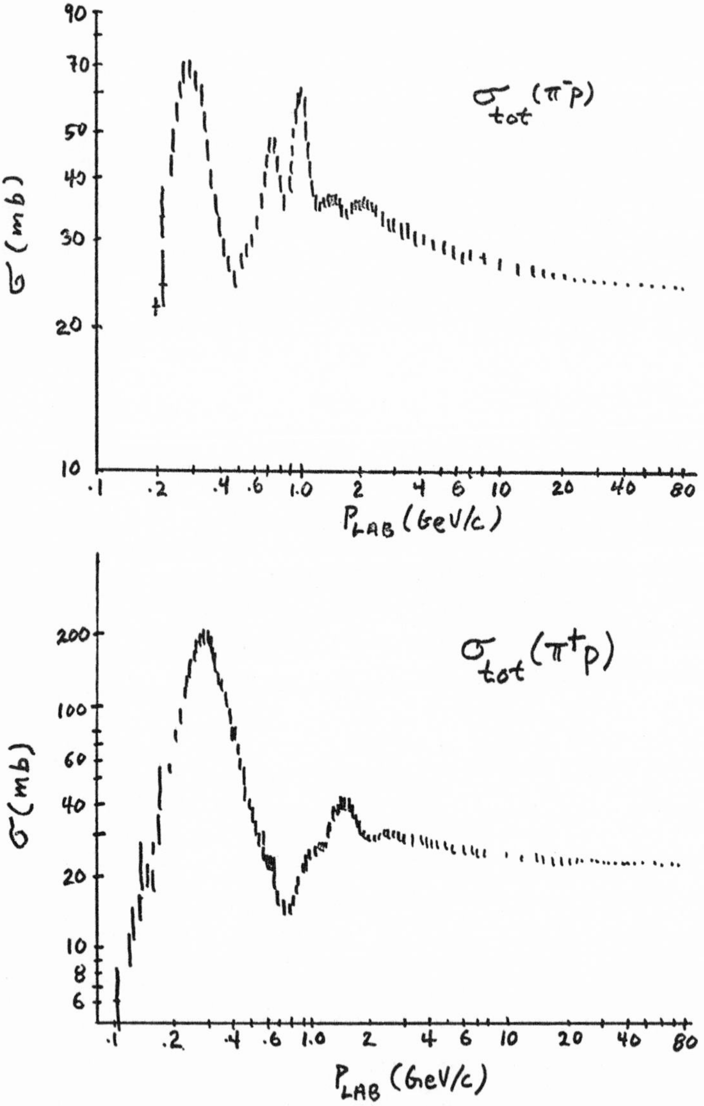

General features of the $\pi^- p$ and $\pi^+ p$ total cross sections, adapted from the compilation by E. Flaminio, J.D. Hauser, D.R.O. Morrison, N. Tovey. CERN-HERA 70-5 and 70-7, 1970.

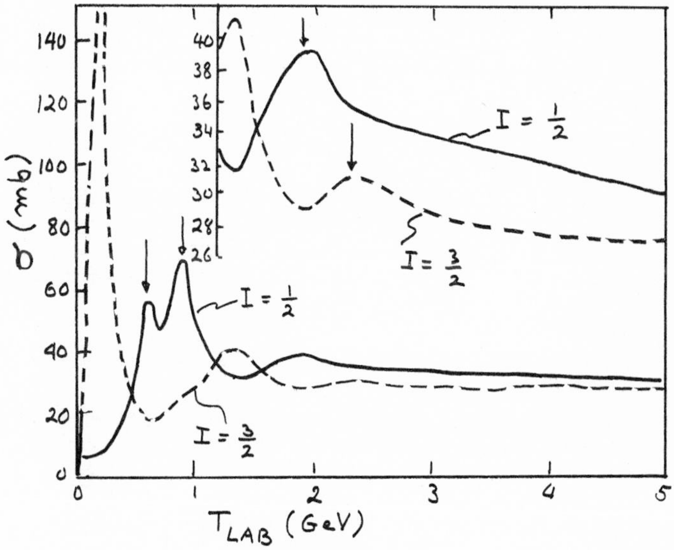

The pion-nucleon total cross section in the I = 1/2 and I = 3/2 states. Adapted from A. Barbaro-Galtieri UCRL-17054, September 30, 1966.

bumps in σ_T often turned out to be due to the superposition of several resonances, each belonging to different partial waves.

Phase Shift Analysis

This will be only a brief mention of how a partial wave analysis is performed in practice. The knowledge of the polarization, total and differential cross sections for the elastic channel, in π^+p and π^-p scattering provides an overconstrained system of equations to be solved for the unknown amplitudes. This system can be solved at each energy, (energy independent partial wave analysis) yielding in general several solutions (e.g., several sets of partial waves which fit equally well the system of equations). Up to this point, all methods are more or less similar. The methods depart from each other when it comes to make a choice of the good solution and to find a continuous behaviour of the partial waves as a function of energy. A typical analysis is for example that of Bareyre et al. (Saclay). The analysis was performed at 13 different energies, between 310 and 990 MeV (T_π).

First step: at energy E

$\frac{d\sigma}{d\Omega}$ 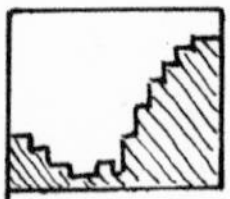 $= |f|^2+|g|^2 = F_1(T_{\ell\pm},\cos\theta)$

$P(\theta)$ 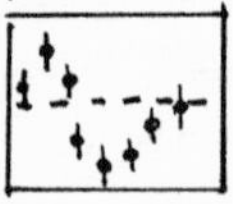 $= \frac{\mathrm{Im}(gh^*)}{|f|^2+|g|^2} = F_2(T_{\ell\pm},\cos\theta)$

} fitted to give $(T_{\ell\pm},\chi^2)_i$

(all i solutions with acceptable χ^2)

<u>Second step</u>: impose continuity on δ and η as a function of energy:

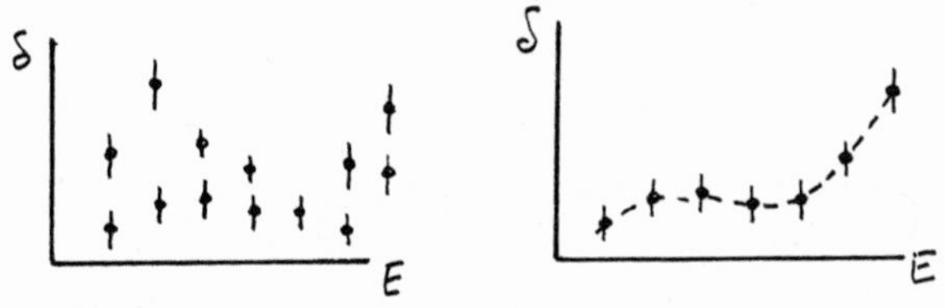

This is done for all partial waves simultaneously.

At each energy the analysis included

$I = 1/2$ S through F = 7 partial waves $14 \times 2 = 28$ parameters to be fitted.
$I = 3/2$ S through F = 7 partial waves ($\uparrow$ η, δ)

With an average of 62 data points per energy. The search for solutions, at each energy, is started from a <u>random set</u> of phase shifts. The argand diagrams of the solution are shown in the accompanying plots.

Resonant behaviour is observed in

S_{11} twice $\to$ N(1530), N(1700) (first loop corresponds to $N\eta$ effect)
P_{11} $\to$ N(1450) (The "Roper resonance")
D_{13} $\to$ N(1527)
D_{15} $\to$ N(1690)
F_{15} $\to$ N(1688)
S_{31} $\to$ Δ(1690)
P_{33} $\to$ Δ(1238) (The well known πN resonance discovered in 1952 by Anderson, Fermi et al.)

Various methods have been devised to make the choice of the best path for each partial wave less arbitrary. The continuity criterion takes the form of the "shortest possible path"

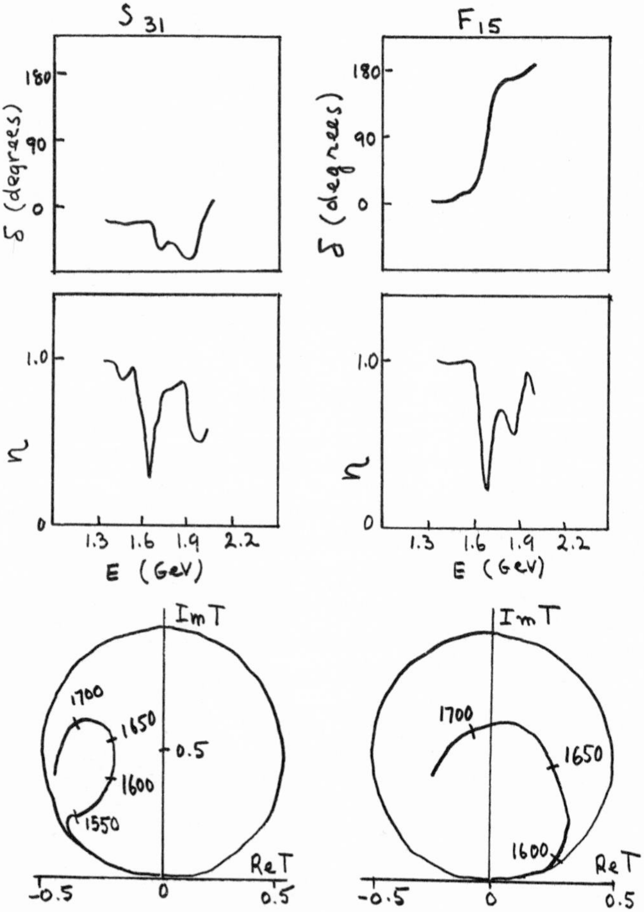

The phase shift δ, the absorption parameter η and the Argand diagrams for the S_{31} and F_{15} partial waves for the πN system, as a typical result of the phase shift analysis of Bareyre et al. (1965).

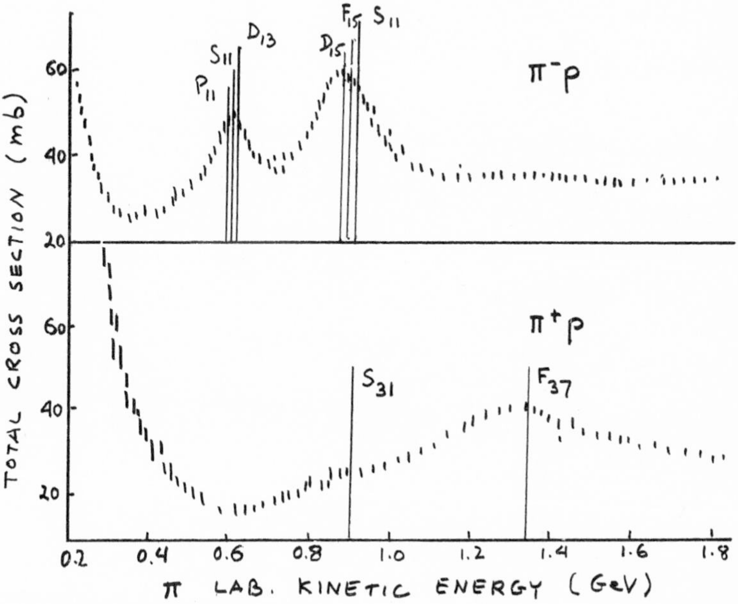

The location of the resonant states detected in the analysis of Bareyre et al. (1965) between 310 and 990 MeV, relative to the behaviour of the π^-p and π^+p total cross sections.

in the complex plane, over a certain energy interval. (Steiner, 1967.) The results turned out very similar to those of Bareyre et al. More sophisticated approaches, making use of dispersion relations to make a choice of the "correct" path for each partial wave, have been used by the CERN group (Donnachie, Kirsopp, Lovelace, 1967).

Here the choice of the "correct" path for each partial wave is made by recourse to theoretical "prejudice," namely by imposing on the real and imaginary parts of the amplitudes, the constraints imposed by <u>dispersion relations</u>. Schematically:

Starting from

$$\mathrm{Re}\left(\frac{T_\ell^\pm}{k}\right) = \frac{P}{\pi}\int_{s_0}^{\infty} \frac{\mathrm{Im}\left(\frac{T_\ell^\pm}{k}\right)}{s'-s}\,ds' + \frac{1}{2\pi i}\int \frac{\Delta\left(\frac{T_\ell^\pm}{k}\right)}{s'-s}\,ds'$$

$s_0 \leftarrow (m_p+m_\pi)^2$; $s \leftarrow E^2$; second integral: over unphysical cuts

one obtains 2 separate relations between Re and Im:

$$\mathrm{Re}\left(\frac{T_\ell^\pm}{k^{2\ell+1}}\right) = \sum_{n=1}^{N} a_n g_n(E) + F_\ell^\pm(E) + \sum_{m=1}^{M} \frac{r_m}{s-b_m}$$

$$\mathrm{Im}\left(\frac{T_\ell^\pm}{k^{2\ell+1}}\right) = \sum_{n=1}^{N} a_n h_n(E)$$

a_n: free parameters

$F_\ell^\pm(E)$: effect of known long range forces

$\frac{r_m}{s-b_m}$: Poles (kept fixed) to approximate short range forces

N,M can be varied.

$T_{\ell}^{\pm}$ are the "experimental" amplitudes.

a_n depend on $\ell^{\pm}$ and can be fitted for each partial wave.

Procedure: DATA

Calculate partial waves using random starting points the first time. Obtain T_{exp} at each energy.

Compare T_{exp} with DR and fit the two, allowing free parameters in DR to move. Obtain T_{corr} and DR_{corr}.

function of energy

If χ^2 is bad, iterate using T_{corr} as starting point. If no convergence is found, increase order of complexity of DR.

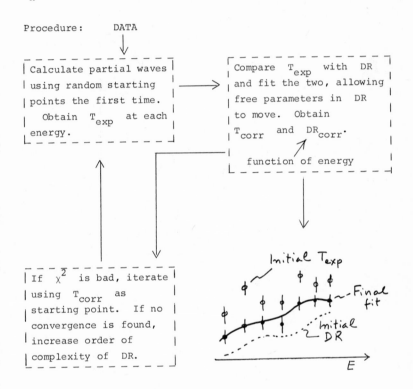

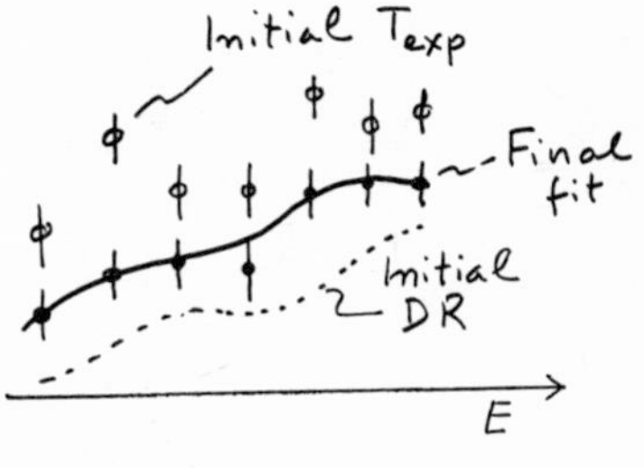

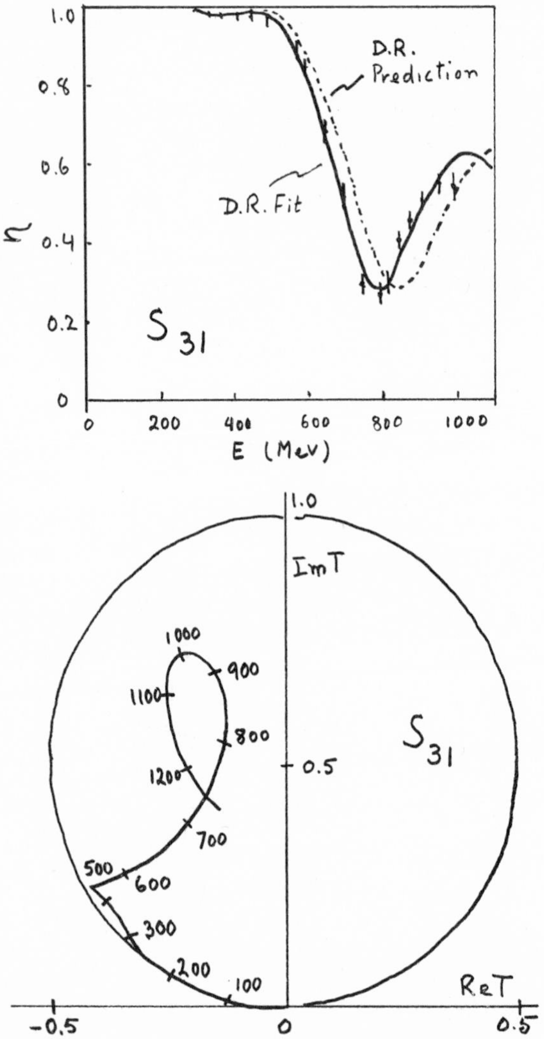

The behaviour of the absorption parameter η and Argand diagram for the S_{31} partial wave, as an example of the results of Donnachie et al. (1965) in the analysis of the πN system.

References

For a comprehensive summary of the latest results concerning resonant states in the pion-nucleon system we refer to Review of Particle Properties. Particle Data Group. Physics Letters 39B, number 1, April 1972, also LBL Report No. 100.

Part of the material in this lecture has been adapted from

M. Ferro-Luzzi in Proceedings of the 1968 CERN School of Physics, El Escorial, 1968. CERN Report 68-23, Vol. II, 1968.

A. Barbaro-Galtieri in Advances in Particle Physics, Vol. 2, edited by R.L. Cool and R.E. Marshak (Wiley & Sons, 1968).

Referred to are papers by

P. Bareyre, C. Bricman, A.V. Stirling and G. Villet, Phys. Letters 18, 342(1965).

H. Steiner, UCRL-17903(1967).

A. Donnachie, A.T. Lea and C. Lovelace, Phys. Letters 19, 146(1965).

20

THE K^- NUCLEON INTERACTION. BARYON RESONANCES WITH STRANGENESS -1.

General Features of $\bar{K}N$ Cross Sections

Characteristic of the $\bar{K}N$ interaction is the presence of several inelastic channels open even at zero energy. Consequently resonant effects may be noticed in the behaviour of the cross sections for all open channels. The isospin decomposition for some of the channels is as follows: since $\bar{K}$ and N are both $I = 1/2$ states

$$|K^-n\rangle = |1\rangle\ ;\ |K^-p\rangle = 1/\sqrt{2}\ |1\rangle + 1/\sqrt{2}|0\rangle\ ;$$
$$|\bar{K}^0 n\rangle = 1/\sqrt{2}|1\rangle - 1/\sqrt{2}|0\rangle \qquad 20.1$$

etc.

Typically, the channels open below $\sim$ 1 Gev/c are

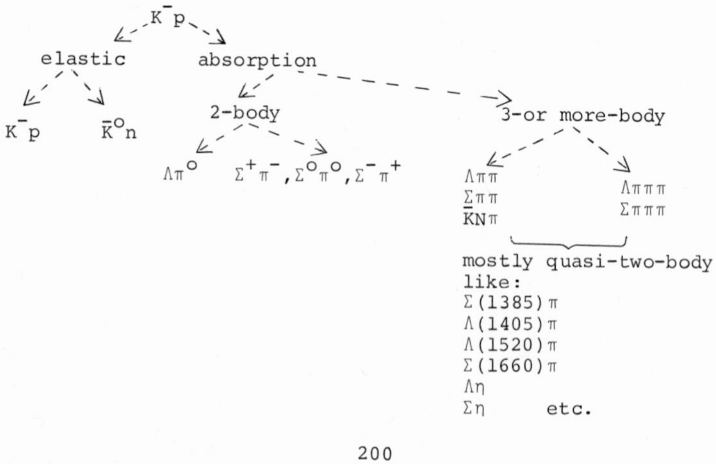

Some of the scattering amplitudes are:

$$T(K^-n \to K^-n) = \langle K^-n | K^-n \rangle = \langle 1|1 \rangle = T_1^{\bar{K}N}$$

$$T(K^-p \to K^-p) = \langle K^-p | K^-p \rangle = 1/2(T_1^{\bar{K}N} + T_0^{\bar{K}N})$$

$$T(K^-p \to \bar{K}^0 n) = \langle \bar{K}^0 n | K^-p \rangle = 1/2(T_1^{\bar{K}N} - T_0^{\bar{K}N}) \qquad 20.2$$

$$T(K^-p \to \Lambda\pi^0) = (\Lambda\pi | K^-p \rangle = 1/\sqrt{2}\ T_1^{\Lambda\pi}$$

$$T(K^-p \to \Sigma^+\pi^-) = \langle \Sigma^+\pi^- | K^-p \rangle = 1/2\ T_1^{\Sigma\pi} + 1/\sqrt{6}\ T_0^{\Sigma\pi}$$

$$T(K^-p \to \Sigma^0\pi^0) = -\ 1/\sqrt{6}\ T_0^{\Sigma\pi}$$

$$T(K^-p \to \Sigma^-\pi^+) = -\ 1/2\ T_1^{\Sigma\pi} + 1/\sqrt{6}\ T_0^{\Sigma\pi}$$

The cross sections of course follow from squaring these amplitudes. Notice that, aside from pure $I = 1$ states, the cross sections will contain interference terms between $I = 0$ and $I = 1$ amplitudes. A trick often used to isolate the contributions of the amplitudes for pure isospin states from the interference terms is that of considering, e.g.,

$$\sigma(K^-p \to K^-p) + \sigma(K^-p \to \bar{K}^0 n) \sim 1/2(|T_1|^2 + |T_0|^2) \qquad 20.3$$

$$\sigma(K^-p \to K^-p) - \sigma(K^-p \to \bar{K}^0 n) \sim \mathrm{Re}\ T_1^* T_0$$

and similarly for the $\Sigma\pi$ system.

For the total cross sections

$$\sigma_T(K^-p) = 1/2(\sigma_1 + \sigma_0); \qquad \sigma_T(K^-n) = \sigma_1 \qquad 20.4$$

Note that while the total K^-p cross section can be measured directly, the K^-n total cross section (and therefore σ_1) can only be obtained indirectly, through scattering of K^- on deuterons.

$$\sigma(K^-d) = "\sigma(K^-p)" + "\sigma(K^-n)" - \sigma_G \qquad 20.5$$

$"\sigma(K^-p)"$ and $"\sigma(K^-n)"$ indicate that the cross sections $\sigma(K^-p)$ and $\sigma(K^-n)$ are smeared by the Fermi motion in the deuteron. σ_G is the so-called Glauber correction, to account for the screening of proton and neutron on each other.

$$\sigma_G \approx 1/4\pi \left\langle r^{-2} \right\rangle \sigma_{(K^-p)} \sigma_{(K^-n)} \qquad 20.6$$

where $\left\langle r^2 \right\rangle$ is the average inverse square separation of the nucleons in the deuteron. The problem of obtaining the pure isospin total cross sections σ_0 and σ_1 is rather laborious. First smooth curves are drawn through the K^-p and K^-d total cross sections. Then $\sigma(K^-p)$ is smeared to simulate the effect of the Fermi motion:

$$"\sigma(K^-p)" = \int |\psi(q)|^2 \sigma(p') d^3q \qquad 20.7$$

where p' is the laboratory momentum corresponding to scattering on a free proton, $\psi(q)$ is the normalized deuteron wave function in momentum space. Then from 20.7, 20.6, 20.5, 20.4, one obtains "smeared" cross sections $"\sigma_0"$ and $"\sigma_1."$ Finally the Fermi momentum is unfolded from $"\sigma_0"$ and $"\sigma_1"$ by an iterative procedure. The enclosed plots show details of the $\bar{K}N$ total cross sections, and the location of several resonances. As in the case of πN scattering, the total cross sections for $\bar{K}N$ gave in several instances the first indication for the existence of a resonance. More often than not, however,

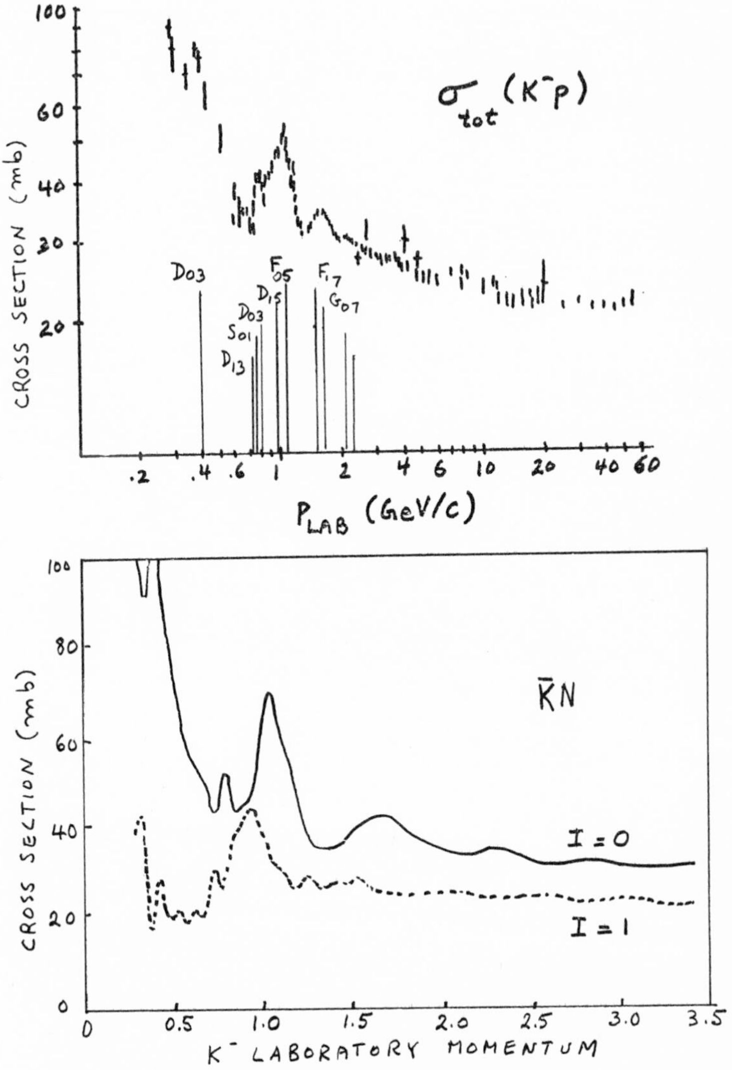

Structure in the K^-p total cross sections, adapted from the CERN-HERA 70-6 computation. The $I = 0,1$ cross sections originate from G. Lynch in Hyperon Resonances-70.

a bump in the total cross section corresponds to the superposition of more than one resonant effect. Detailed study of all channels are required to resolve such structures, and this has been accomplished to a large extent by experiments in the H_2 and D_2 bubble chamber.

$\Lambda(1520)$ and $K^-\Sigma$ Relative Parity

The study of this resonance is in a sense a classic example, such as the study of $\Delta(1236)$ in the πN interaction. $\Lambda(1520)$ is the first state above $\bar{K}N$ threshold. It occurs at $\sim$ 390 Mev/c K^- lab. momentum. It was discovered and studied in detail in a formation experiment by Watson, Ferro-Luzzi and Tripp in 1962-63 (Phys. Rev. 131, 2248 (1963)). All the conclusions regarding the quantum numbers of this state had already been derived at that time. The same experiment has been recently repeated at LBL, with 100 times the old statistics! Will briefly review some of the old evidence, illustrated now also by the much better data recently become available. The behaviour of the cross sections for some of the most important channels, is shown in the accompanying graphs. Evidence for resonant behaviour at $\sim$ 390 Mev/c is observed in all channels, except $\Lambda\pi^o$. The quantum numbers of this state can be derived from simple arguments, although much stronger conclusions have in fact been derived from a multi-channel K-matrix analysis of the data.

I spin: a) No enhancement in $\Lambda\pi$ $(I = 1)$

b) Strong enhancement in $\Sigma^o\pi^o$ $(I = 0)$

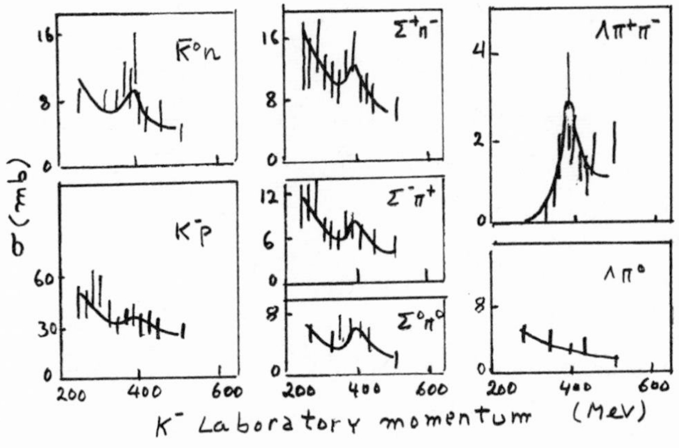

Study of Λ(1520) formation, adapted from M.B. Watson, M. Ferro-Luzzi and R.D. Tripp (1963).

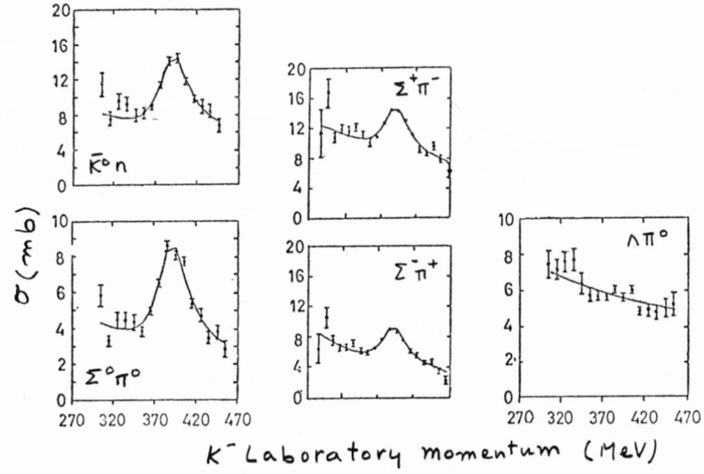

Improved study of Λ(1520) formation, by A. Barbaro-Galtieri, M. Alston-Garnjost, R.D. Bangerter, L.K. Gershwin, T. Mast and R.D. Tripp, as reported at the Lund Conference (1969).

c) Branching ratio $\Lambda\pi^2\pi^0 / \Lambda\pi^+\pi^- \approx 1/2$ (as expected for $I = 0$, rather than zero for $I = 1$).

d) Observed in production experiments only in neutral states.

Conclusion: $\underline{I = 0}$.

Spin: follows from the behaviour of the angular distributions. These show a strong $\cos^2\theta$ dependence in the resonance region and no $\cos^4\theta$ dependence, indicating $J > 1/2$ and $< 5/2$, most probably 3/2. In terms of Legendre coefficients, no term higher than A_3 is required in any of the channels. A strong peak is observed in A_2 around 390 MeV/c. Since

$$A_0 = |S_1|^2 + |P_1|^2 + 2|P_3|^2 + 2|D_3|^2$$

$$A_1 = 2S_1^*P_1 + 4(S_1^*P_3 + P_1^*D_3) + 4/5\, P_3^*D_3$$

$$A_2 = 4(S_1^*D_3 + P_1^*P_3) + 2|P_3|^2 + 2|D_3|^2$$

Partial waves through $J = 3/2$ only.

20.8

$$A_3 = 36/5\, P_3^*D_3 \quad (\text{read } \mathrm{Re}\, P_3^*D_3 \text{ for } P_3^*D_3)$$

the candidates for explaining this effect are clearly the P_3 and D_3 partial wave. ($3/2^+$ or $3/2^-$).

Parity: the S_1 partial wave is known to be strong from threshold up to 300-400 MeV/c. If the resonance were in a P_3 state (in the $\bar{K}N$ channel), the term $S_1^*P_3$ in A_1, would give a major contribution in A_1, contrary to the evidence. A strong contribution to A_2 from the $S_1^*D_3$ term is on the other hand in agreement with all observations. $J^P = 3/2^-$. (Notice that this is only a qualitative argument.)

<u>Branching Ratios</u>: These quantities have important implications related to SU(3) as we shall see. For $\Lambda(1520)$ they have been the subject of debate for many years, in fact the original branching ratios of W.F.T. were in disagreement with those found in production experiments. The new LBL experiment on the other hand, due in particular to the better resolution of the $\bar{K}^0 n$ peak, agrees with the results of production experiments. The <u>spin</u> of $\Lambda(1520)$ can be deduced from <u>unitarity</u>, using the new data, and

$$\Gamma = \underbrace{\Gamma_{\bar{K}N}}_{\Gamma_e} + \underbrace{\Gamma_{\Sigma\pi} + \Gamma_{\Lambda\pi\pi} + \Gamma_{\Sigma\gamma} + \underbrace{\Gamma_{\Lambda\gamma} + \Gamma_{\Sigma\pi\pi}}_{\text{small}}}_{\Gamma_r} \qquad 20.9$$

The elasticity $\Gamma e/\Gamma$ is easily obtained from the size of the resonant bump in $\bar{K}p \to \bar{K}^0 n$.

$$\sigma_{\bar{K}^0 n} = \underset{\text{from isospin}}{1/4}\; 4\pi\lambda\!\!\!^{-2}(J+1/2)x^2 = \sim 8.4\text{ mb} \begin{cases} x = 0.45 & \text{if } J = 3/2 \\ x = 0.36 & \text{if } J = 5/2 \end{cases}$$

$$\sigma_r = \underset{\text{from isospin}}{1/2}\; 4\pi\lambda\!\!\!^{-2}(J+1/2)x(1-x) \begin{cases} 20.8\text{ mb} & \text{if } J = 3/2 \\ 29\text{ mb} & \text{if } J = 5/2 \end{cases}$$

By adding up the sizes of the resonant bumps in all inelastic channels, $\sigma_r = (20.9 \pm 0.4)$mb which clearly selects $J = 3/2$ from unitarity.

A new precise determination of the branching fractions in the decay of $\Lambda(1520)$ as observed in <u>production</u> has been obtained by a Chicago-CERN-Heidelberg-Saclay collaborative experiment, where $\Lambda(1520)$ was observed in the chain decay

$$K^-p \to \Lambda(1760) \to \Lambda(1520\ \pi^0 \qquad 20.10$$

$\Lambda(1520)$ Decay Branching Fractions (%)

Mode	Formation (LBL)	Production (CCHS)
$\bar{K}N$	44.5 ± 1.8	47.2 ± 2.2
$\Sigma\pi$	41.7 ± 1.7	38.6 ± 2.6
$\Lambda\pi\pi$	10.4 ± 0.6	10.3 ± 1.4
$\Sigma\gamma$	1.8 ± 0.3	
$\Lambda\gamma$	0.7 ± 0.2	
$\Sigma\pi\pi$	0.9 ± 0.2	

20.11

$\bar{K}$ Relative Parity. The W.F.T. experiment gave also the first direct determination of the $\bar{K}\Sigma$ parity, based on the behaviour of the polarization the the $\Sigma\pi$ channel. See original experiment or D.H. Miller, p. 66. Briefly, the polarization angular distribution is observed as predicted from dominant $S_1^*D_3$ interference ($\bar{K}N\Sigma$ parity odd). For $\bar{K}N$ parity even, the dominant term would be $P_1^*P_3$, of sign opposite to that of $S_1^*D_3$.

$$I(\theta)\vec{P}(\theta) = \hat{n}\lambda\!\!\!^{-2}\Sigma B_n P_n^1(\cos\theta)$$

$$B_1 = 2S_1^*P_1 + 1.6P_3^*D_3$$

$$B_2 = 2S_1^*D_3 - 2P_1^*P_3$$

$$(S_1^*D_3 \equiv \mathrm{Im}S_1^*D_3) \text{ etc.} \qquad 20.12$$

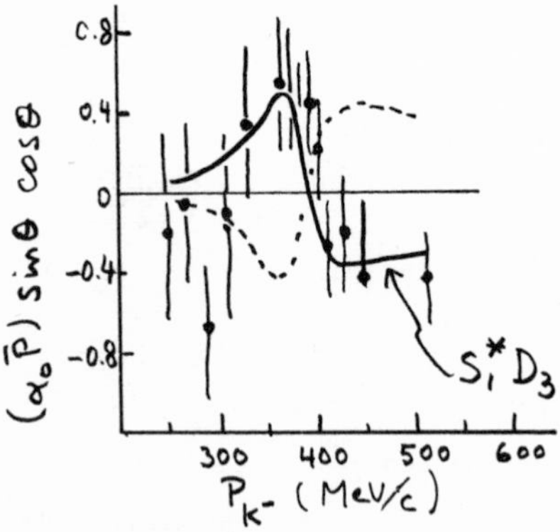

B_2 gives $\sin\theta\ \cos\theta$ terms in angular distribution.

(Note that the Wigner condition removes the complex conjugation ambiguity. Must take D_3 and not D_3^*.)

Resonances in the 1 Gev/c Region

Situation as of 1963:

a) From $\sigma_{tot}(K^-p)$, "Kerth bump" at ~ 1.05 Gev/c, $\rightarrow$ evidence for $\Lambda(1815)$. $I = 0$ from apparent absence of a peak twice as large in K^-n. In fact from

$$\begin{aligned} \sigma_0(\bar{K}N) &= 2\sigma(K^-P) - \sigma(K^-n), \\ \sigma(K^-n) &= 2\sigma(K^-p) \quad \text{if} \quad I = 1. \end{aligned} \qquad 20.13$$

b) Angular distributions in $K^-p \rightarrow K^-p$, $\bar{K}^on$ have complexity requiring A_5. Thus $J = 5/2$ most likely.

c) Evidence for a resonance at $M = 1765$ Mev found by Barbaro-Galtieri, Hussain and Tripp in the (K^-p) invariant mass spectrum from $K^-n \rightarrow K^-p\pi^-$ at 1.51 Gev/c. Speculation of $I = 1$ for the new effect based on the following argument: A_5 positive for $K^-p \rightarrow K^-p$, negative for $K^-p \rightarrow \bar{K}^on$. In fact $A_5 = 14.3\, \mathrm{Re}(D_5^*F_5)$ if terms of $J \geq 7/2$ are neglected.

From the amplitudes

$$\begin{aligned} A_5(\bar{K}^on) &= \frac{14.3}{4}\, \mathrm{Re}\left[(D_{15}-D_{05})^*(F_{15}-F_{05})\right] \\ A_5(K^-p) &= \frac{14.3}{4}\, \mathrm{Re}\left[(D_{15}+D_{05})^*(F_{15}+F_{05})\right] \end{aligned} \qquad 20.14$$

The assumption of resonant $F_{05}(1815)$ and $D_{15}(1765)$, with small D_{05} and F_{15} leads to A_5 of opposite sign.

These conjectures were soon confirmed in a systematic exploration of the 500-1200 Mev/c region, by CERN-Heidelberg-Saclay, joined by Chicago for the study of $K^-p \rightarrow K^-p$ and $K^-p \rightarrow \bar{K}N\pi$.

Among the first results of the CCHS experiment was the discovery of the chain decay $K^-p \to \Sigma(1765) \to \Lambda(1520)\pi^0$ which led to an unambiguous assignment $I = 1,\ J^P = 5/2^-$ to $\Sigma(1765)$. (See Lecture 21.)

After six years of formation experiments by several laboratories, various partial wave analyses have been performed, on most of the elastic and inelastic channels. Only in a few cases (like $K^-p \to \Sigma\pi$) where both $d\sigma/d\Omega$ and $P(\theta)$ data were available, an <u>energy-independent</u> phase shift analysis, of the type described for the πN interaction, could be performed. Most analyses have been <u>energy dependent</u>, namely the behaviour of each partial wave was fitted to the data, for a particular model of its energy dependence. Example:

Non-resonant partial waves: $T = a + b(P_K - 1)$ a, b_1 complex.

Resonant p.w. : $T = \dfrac{x}{\varepsilon - i}$ 20.15

Both R + NR p.w. : $T = T_b + T_r\, e^{2i\delta}$

Several more resonant states were discovered by this approach in the 1 Gev/c region:

$\Lambda(1700),\ J^P = 3/2^-$ (observed first in σ_{tot} at RHEL)

$\Lambda(1830),\ J^P = 5/2^-$ in addition to

$\Lambda(1870),\ J^P = 3/2^+$ $\Sigma(1765),\ J^P = 5/2^-$

$\Sigma(1760),\ J^P = 1/2^-$ $\Lambda(1815),\ J^P = 5/2^+$

Only a few of the characteristic features of the experimental data and the behavior of the lower partial waves for, e.g., the $\Sigma\pi$ channel are illustrated in the following pages.

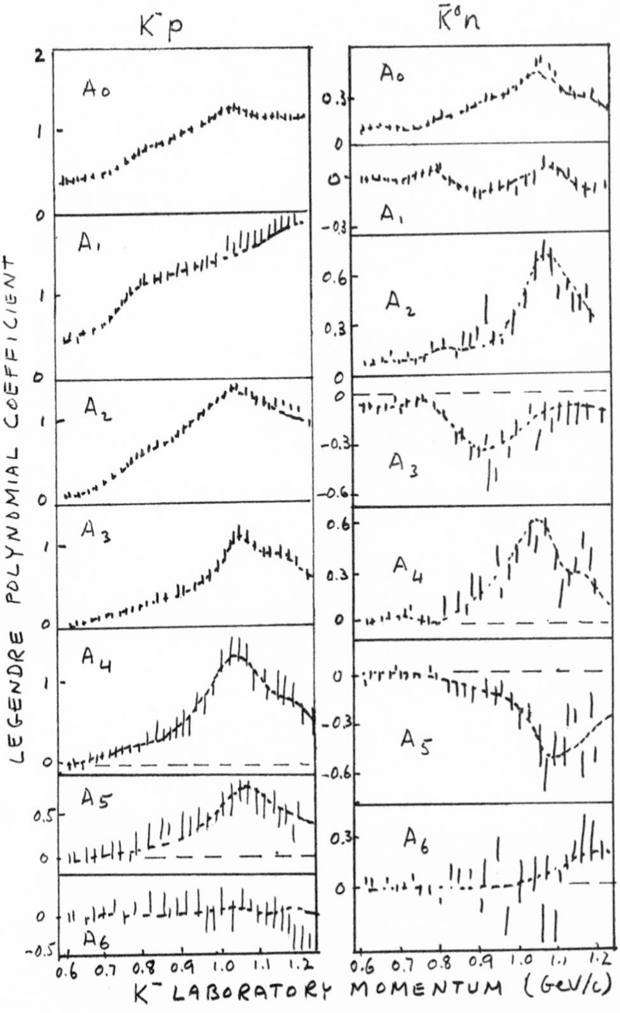

The behaviour of the Legendre Polynomial Coefficients in the expansion $\frac{d\sigma}{d\Omega} = \bar{\lambda}\sum A_n P_n(\cos\theta)$, as a function of K^- laboratory momentum, for $K^-p \rightarrow K^-p$ and $K^-p \rightarrow \bar{K}^0 n$. Adapted from R. Armenteros et al. Nuclear Physics B8, 195(1968).

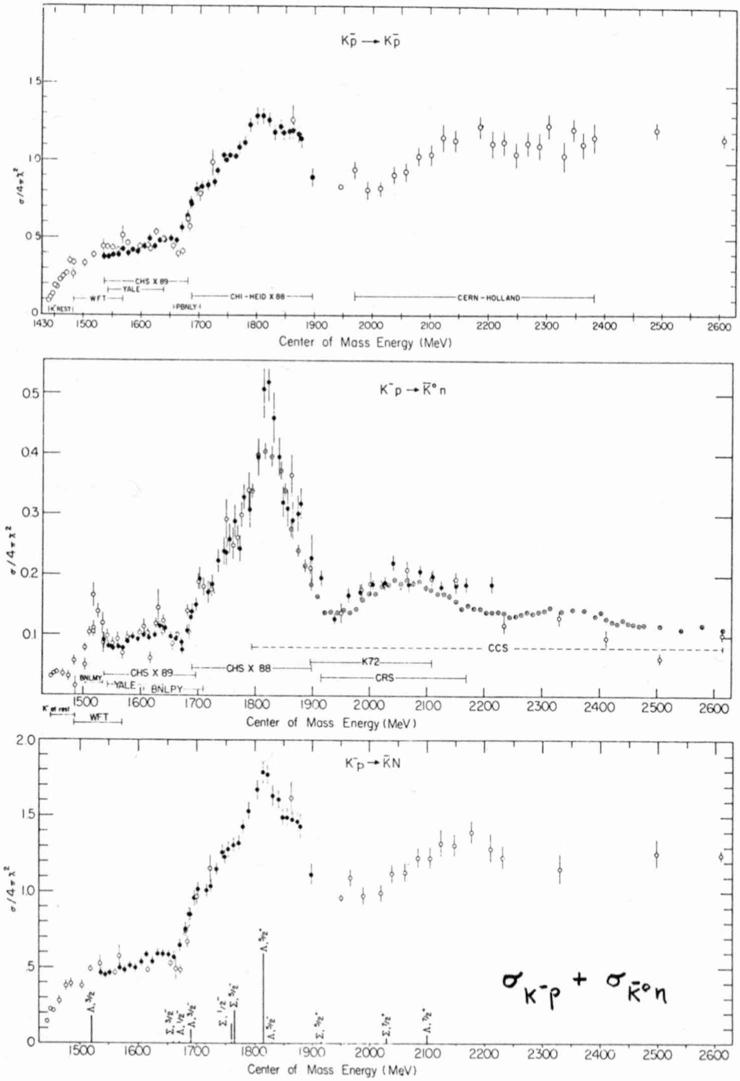

Structure in the elastic channel cross sections, for the K^-p interaction.

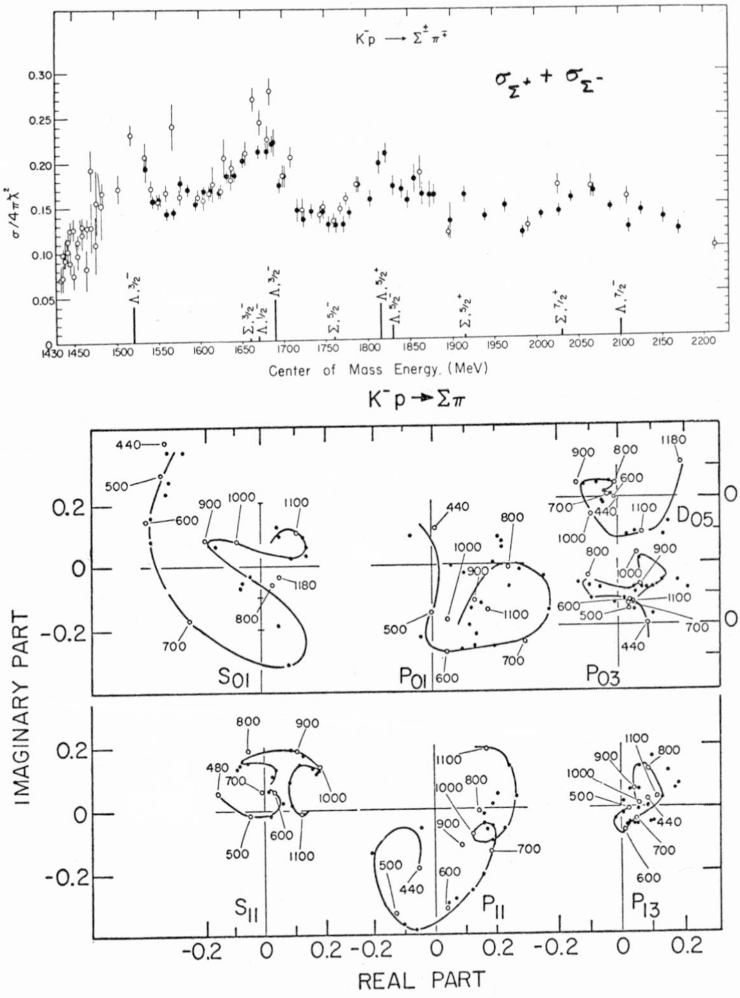

Structure in the $K^-p \rightarrow \Sigma\pi$ channel. The Argand diagrams of the lower partial waves were obtained in an energy-independent partial wave analysis by the CERN-Heidelberg-Saclay groups at the time of the Lund Conference (1969).

References

Comprehensive reviews of this topic can be found in:

A. Barbaro-Galtieri in Advances in Particle Physics, Vol. 2, edited by R.L. Cool and R.E. Marshak (Wiley & Sons, 1968).

D.H. Miller in High Energy Physics, Vol. IV, edited by E.H.S. Burhop, Academic Press, New York & London, 1969.

Hyperon Resonances-70, edited by E.C. Fowler, Moore Publishing Co., Durham, North Carolina.

R. Levi Setti in Proceedings of the Lund International Conference on Elementary Particles. Edited by G. von Dardel, Berlingska Boktryckeriet, Lund, Sweden, 1969.

21

SEQUENTIAL DECAY OF RESONANT STATES

We will consider in some detail a specific example of sequential decay of a hyperon resonance, $\Sigma(1760)$, to illustrate a very general method for the determination of spin and parity of resonant states. In fact the example to be discussed encompasses methods appropriate to the analysis of <u>formation</u> and <u>production</u> experiments at the same time. The sequential decay in question originates in the following reaction

$$K^- + p \to \Sigma(1760) \to \Lambda(1520)\pi^0$$

$$\quad \to K^- p$$

$$\quad \to \Sigma\pi$$

$$\quad \to \Lambda\pi\pi$$

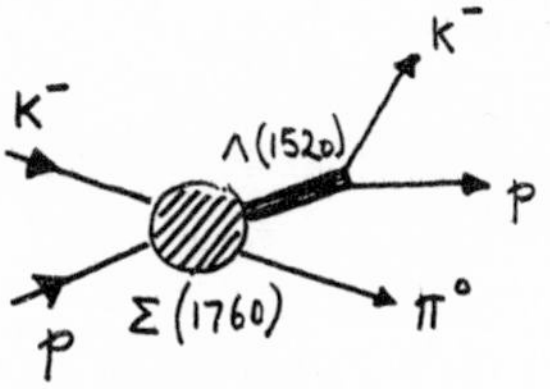

21.1

Here we have the <u>formation</u> of $\Sigma(1760)$ in K^-p interaction around 1760 MeV c.m. energy, followed by the <u>production</u> of $\Lambda(1520)$ as a final product of the decay of $\Sigma(1760)$. This chain was discovered during a systematic investigation of the K^-p interaction in the 1 GeV/c region, by a collaboration of CERN, Chicago, Heidelberg, Saclay and independently also at Berkeley. At the time of its discovery, the basic unknown in 21.1 was the J^P of $\Sigma(1760)$, although the very existence of this resonant state was still in need of confirmation. The nature of the evidence for reaction 21.1 is the following:

a) Copious production of $\Lambda(1520)$ is observed in the channels $\Sigma^{\pm}\pi^{\mp}\pi^{0}$, $\Lambda\pi^{\pm}\pi^{\mp}\pi^{0}$ and $K^{-}p\pi^{0}$ in a region around 930 MeV/c lab. momentum of the incident K^{-}:

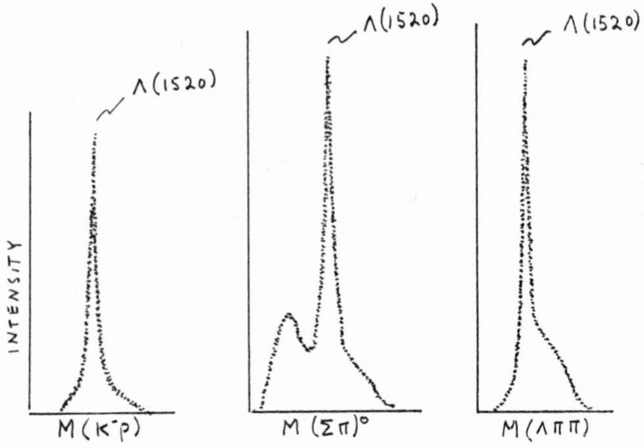

b) The cross section for the production of $\Lambda(1520)$, when plotted as a function of, e.g., K^{-} incident momentum, shows a clear enhancement, which actually represents the Breit-Wigner of $\Sigma(1760)$.

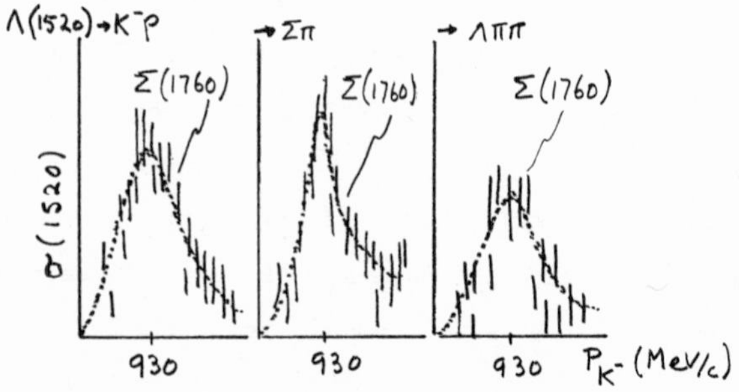

Several of the quantum numbers of $\Sigma(1760)$ can be derived from just a) and b). In fact

$$I = 1 \quad \text{from} \qquad \Sigma(1760) \rightarrow \Lambda(1520) + \pi^{0}$$

$$I = \quad 1 \qquad\qquad 0 \qquad 1$$

The mass M and width Γ are determined by a fit to the above production cross sections.

For the J^P determination we illustrate here the method followed by Fenster et al. P. R. L. 17, 841(1966) which examines the joint correlation of the production and decay distributions of $\Lambda(1520)$. This is nothing else but an application of conservation of angular momentum and parity. It is however not trivial. What do we know about the angular momenta involved in 21.1? The relevant choices are:

$$K^- + p \rightarrow \Sigma(1760) \rightarrow \Lambda(1520) + \pi^{o} \tag{21.2}$$

$$J^P = \tfrac{5}{2}^+ \text{ or } \tfrac{5}{2}^- \qquad \tfrac{3}{2}^- \qquad 0^-$$

$\tfrac{5}{2}^-$: $\ell = 1, 3 \ldots$

$\tfrac{5}{2}^+$: $\ell = 2, 4 \ldots$

Rather than the ℓ, s representation we will however work in the helicity representation (M. Jacob and G. C. Wick, Annals of Physics: 7, 404(1959).). For additional details on the specific problem at hand, see also J. D. Jackson, High Energy Physics, Les Houches (1965), edited by De Witt & Jacob.

Consider first the decay of a particle γ with spin j and projection m

$$\gamma \to \alpha + \beta \qquad 21.3$$

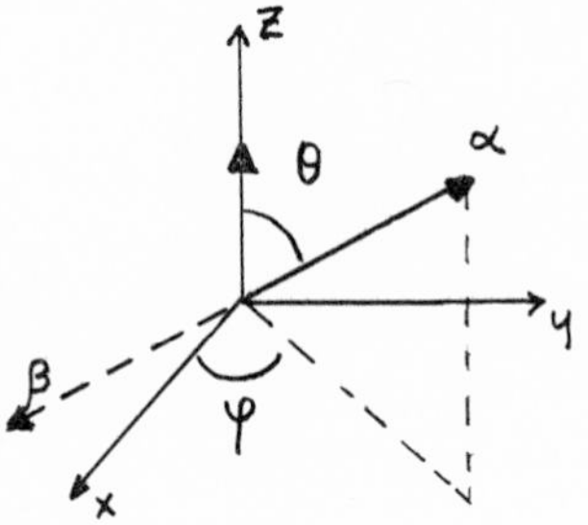

where α and β have helicities λ_α, λ_β. (The λ's are the eigenvalues of the helicity operator $\Lambda = \frac{\vec{J}\cdot\vec{P}}{|\vec{p}|} = \vec{J}\cdot\hat{p}$).
The amplitude for the decay is given by

$$A_m^{(\lambda_\alpha\lambda_\beta)} = \sqrt{\frac{2j+1}{4\pi}}\, a(\lambda_\alpha,\lambda_\beta)\mathcal{D}^{j*}_{m\lambda}(\phi,\theta,0) \qquad 21.4$$

where $\lambda = \lambda_\alpha - \lambda_\beta$ and $a(\lambda_\alpha,\lambda_\beta)$ is the matrix element of a transition operator U in the helicity frame. Will specify this later. The $\mathcal{D}$ functions are the usual representations of the rotation group. Here θ,ϕ are the decay angles of α,β relative to the helicity axis of γ. That is, γ was moving in some direction which defined its helicity direction. We have taken this to coincide with our z-axis. θ and ϕ refer to the rest frame of γ which is reached by a Lorentz transformation along its helicity direction. Expression 21.4 will be needed for the construction of the density matrix for the decay of $\Sigma(1760)$.

In the full chain 21.1 we will have to consider two density matrices:

$$K^-p \to \Sigma(1760) \to \Lambda(1520) + \pi \qquad 21.5$$

(1) (2)

$\rho_{\mu\mu'}$ j_1 $\rho_{\lambda\lambda'}$ j_2

In the formation of $\Sigma(1760)$, the latter appears as an <u>aligned</u> state. In the overall C.M. system, the only populated states will be those with $M_j = \pm 1/2$, since only

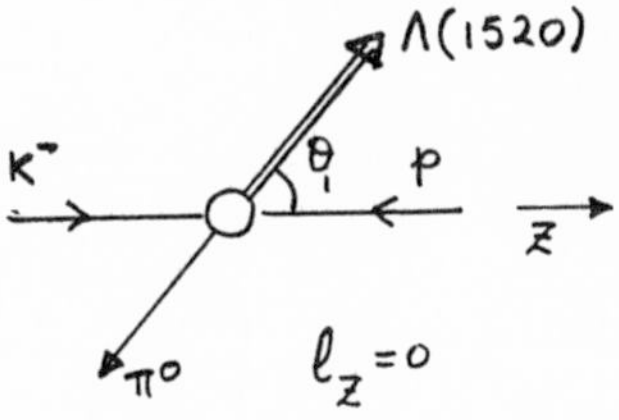

the proton has spin, $J_p = 1/2$. Thus the total helicity of the initial two particle system is $\pm 1/2$. Since $\ell_z = 0$, there can be no ϕ_1 dependence in $\Lambda(1520)$ production. We may therefore set $\phi_1 = 0$ (definition of $\theta_1\phi_1$ coordinates for <u>each</u> event!). If we make the dynamical assumption that all $K^-p\pi^0$ comes from the sequential chain, the density matrix for the <u>formation</u> of $\Sigma(1760)$ is then

$$\rho^{(1)}_{\mu\mu'} = \begin{pmatrix} & & \vdots & \vdots & & \\ & & \vdots & \vdots & & \\ & & 1/2 & 0 & \cdots & \\ & & 0 & 1/2 & \cdots & \\ & & & & & \\ & & & & & \end{pmatrix} \quad \begin{matrix} 5/2 \\ 3/2 \\ 1/2 \\ -1/2 \\ -3/2 \\ -5/2 \end{matrix} \qquad 21.6$$

(columns: 5/2, 3/2, 1/2, −1/2, −3/2, −5/2)

or

$$\rho^{(1)}_{\mu\mu'} = 1/2\ \delta_{\mu\mu'}(\delta_{\mu 1/2}+\delta_{\mu -1/2}) \qquad 21.7$$

The $\Lambda(1520)$ will be produced in the spin states described by

$$\rho^{(2)}_{\lambda\lambda'}(\theta_1,0) = \Sigma_{\mu\mu'} A_\mu(\lambda_{1520},\lambda_\pi)\rho^{(1)}_{\mu\mu'} A^*_{\mu'}(\lambda_{1520},\lambda_\pi) \qquad 21.8$$

where A_μ is the production amplitude of $\Lambda(1520)$ or the decay amplitude of $\Sigma(1760)$. From 21.4 we have

$A_\mu(\lambda_{1520},\lambda_\pi) = a(\lambda)\mathcal{D}^{5/2^*}_{\mu\lambda}(0,\theta_1,0)$ 21.9 since we have set $\phi_1 = 0$, neglecting the factor $\sqrt{(2J+1)/4\pi}$ which will be included in an overall normalization factor. Since $\lambda_\pi = 0$, the amplitude $a(\lambda)$ is a function only of the helicity of $\Lambda(1520)$, $\lambda_{1520} = \lambda = \pm 3/2, \pm 1/2$. Important: there are parity constraints on $a(\lambda)$, such that

$$a(-\lambda) = \eta_{1760}\eta_{1520}\eta_{\pi^0}(-1)a(\lambda) \rightarrow a(-\lambda) = -\eta_{1760}a(\lambda) \qquad 21.10$$

This relation will provide a very sensitive test of the parity of $\Sigma(1760)$.

At this point one could calculate the production angular distributions of $\Lambda(1520)$, $I(\theta_1,\phi_1)$, by taking the trace of $\rho^{(2)}_{\lambda\lambda'}$. We go on instead to calculate the joint correlation function.

To describe the $\Lambda(1520)$ decay we go to its rest system and define a new axis of quantization $\hat{z}_2$ along its line of motion, as seen from the θ_1,ϕ_1 system. This new axis of quantization is still in the production plane (like $\hat{z}_1$) but is rotated by θ_1,ϕ_1 (or $\theta_1,0$) in our case. θ_2,ϕ_2 will be measured with respect to this new axis, for the time being. The joint correlation distribution is given by, in general

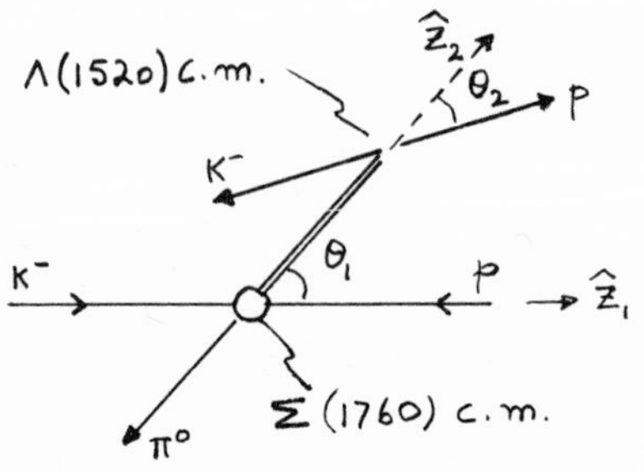

$$I(\theta_2\phi_2;\theta_1\phi_1) = \sum_{\substack{\lambda_\alpha\lambda_\beta\\ \lambda\lambda'}} |M(\lambda_\alpha\lambda_\beta)|^2 \mathcal{D}^{j_2 *}_{\lambda\lambda_2}(\theta_2\phi_2)\mathcal{D}^{j_2}_{\lambda'\lambda_2}(\theta_2\phi_2)\rho^{(2)}_{\lambda\lambda'}(\theta_1\phi_1) \qquad 21.11$$

(Jackson 3.11).

Here

$$j_1 = 5/2,\ j_2 = 3/2,\ \lambda_2 = \lambda_\alpha - \lambda_\beta; \quad \lambda_\beta = 0 \quad (K^-)$$
$$\lambda_\alpha = \pm 1/2 \text{ (proton)}$$

$M(\lambda_\alpha, \lambda_\beta) \to M(\pm 1/2)$ is the amplitude for the decay $\Lambda(1520) \to K^- p$. Since it appears squared, we can absorb it in an overall normalization. With the choice $\phi_1 = 0$

$$(\Lambda(1520) \text{ production plane} \equiv x - z_1 \text{ plane})$$

we have

$$\rho_{mm'} = (-1)^{m-m'} \rho_{-m-m'} \quad 21.12 \quad \text{(Jackson 3.20)}.$$

Then
$$I(\phi_2 \theta_2; \theta_1, 0) = N\{(1/3 + \cos^2\theta_2)\rho^{(2)}_{11}(\theta_1) + \sin^2\theta_2 \rho^{(2)}_{33}(\theta_1)$$
$$-2/\sqrt{3}\cos\phi_2 \sin 2\theta_2 \operatorname{Re}(\rho_{31}(\theta_1)) \qquad 21.13$$
$$-2/\sqrt{3}\cos 2\phi_2 \sin^2\theta_2 \operatorname{Re}(\rho^{(2)}_{3-1}(\theta_1))$$

(Jackson 3.19).

In the analysis of Fenster et al. this expression was integrated over ϕ_2 (due to poor statistics):

$$I(\theta_2, \theta_1) = N\{\rho^{(2)}_{33}(\theta_1) + \rho^{(2)}_{11}(\theta_1) + \left[\rho^{(2)}_{11}(\theta_1) - \rho^{(2)}_{33}(\theta_1)\right] P_2(\cos\theta_2)\} \qquad 21.14$$

↑ Overall normalization to number of events (refers to N)

where we have used $\cos^2\theta_2 = 2/3 P_2(\cos\theta_2) + 1/3$.

There is another choice for the decay angle of $\Lambda(1520)$ which turned out to be particularly sensitive to the $\Sigma(1760)$ parity. It consists in measuring the decay angle relative to an axis $\hat{z}_3$ normal to the production plane. 21.14 will still be valid, provided we rotate $\rho^{(2)}$ by $\pi/2$.

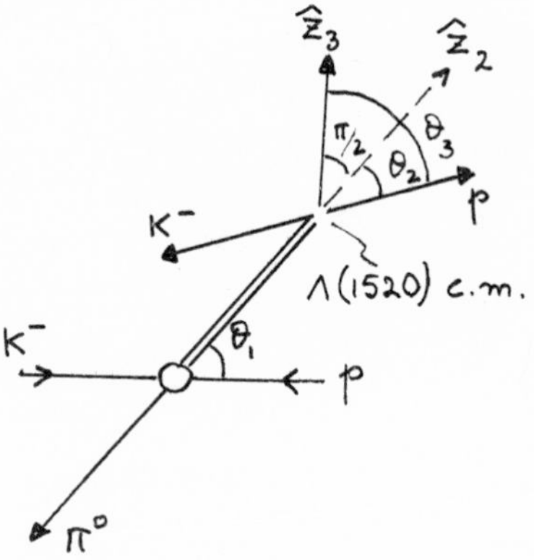

$$I(\theta_3,\theta_1) = N\{(\rho^{(2)}_{33}(\theta_1))_R + (\rho^{(2)}_{11}(\theta_1))_R + \left[(\rho^{(2)}_{11}(\theta_1))_R - (\rho^{(2)}_{33}(\theta_1))_R\right]P_2(\cos\theta_3)\} \quad 21.15$$

The rotation of the density matrix is obtained from

$$(\rho^{(2)}_{\mu'\mu})_R = (\mathcal{D}^{3/2}\rho^{(2)}\mathcal{D}^{3/2^{-1}})_{\mu'\mu} = \sum_{\eta\eta'} d^{3/2}_{\mu'\eta}(\pi/2)\,\rho^{(2)}_{\eta\eta'}\,d^{3/2}_{\mu\eta'}(\pi/2)\,e^{i(\eta'-\eta)\pi/2} \quad 21.16$$

(cf. M.E. Rose, Elementary Theory of Angular Momentum, John Wiley & Sons, 1957.)

From Berman and Jacob, P.R. 139, B1023 (1965),

$$d^{3/2}_{\mu\mu'}(\pi/2) = 1/\sqrt{8}\begin{pmatrix} 1 & -\sqrt{3} & \sqrt{3} & -1 \\ \sqrt{3} & -1 & -1 & \sqrt{3} \\ \sqrt{3} & 1 & -1 & -\sqrt{3} \\ 1 & \sqrt{3} & \sqrt{3} & 1 \end{pmatrix} \quad 21.17$$

From 21.15, 21.16, 21.17 and 21.12 one finds:

$$I(\theta_3,\theta_1) = N\{\rho^{(2)}_{33}(\theta_1) + \rho^{(2)}_{11}(\theta_1) + 1/2\Big[(\rho^{(2)}_{33}(\theta_1)-\rho^{(2)}_{11}(\theta_1)) + \sqrt{3}(\rho^{(2)}_{3-1}(\theta_1) + \rho^{(2)}_{-13}(\theta_1))\Big]P_2(\cos\theta_3)\} \qquad 21.18$$

Notice that the coefficient of $P_2(\cos\theta_3)$ in 21.18 is $-1/2$ that of $P_2(\cos\theta_2)$ in Equation 21.14 plus terms $\propto a(3/2)a^*(-1/2) + a(-1/2)a^*(3/2)$ (from 21.8 and 21.9). Because of 21.10, the coefficient of $P_2(\cos\theta_3)$ is sensitive to η_{1760}.

We still have to evaluate $\rho^{(2)}_{\lambda'\lambda}(\theta_1)$ explicitly. From 21.6, 21.8 and 21.9 the explicit form of this density matrix is:

$$\rho^{(2)}_{\lambda'\lambda}(\theta_1) = 1/2a(\lambda)a^*(\lambda')\{\mathcal{D}^{*5/2}_{1/2\lambda'}\mathcal{D}^{5/2}_{1/2\lambda} + \mathcal{D}^{*5/2}_{-1/2\lambda'}\mathcal{D}^{5/2}_{-1/2\lambda}\} \qquad 21.19$$

Using now properties and relations for the $\mathcal{D}$ functions (see Rose)

$$\mathcal{D}^{j*}_{m'm} = (-1)^{m'm}\mathcal{D}^{j}_{-m'-m} \qquad 21.20$$

$$\mathcal{D}^{j_1}_{\mu_1 m_1}\mathcal{D}^{j_2}_{\mu_2 m_2} \qquad 21.21$$

$$= \sum_j (j_1\mu_1+\mu_2|j_1,j_{2i}\mu_1,\mu_2)(j,m_1+m_2|j_1,j_2;m_1,m_2)\mathcal{D}^{j}_{\mu_1+\mu_2,m_1+m_2}$$

In our case

$$\mathcal{D}^{5/2*}_{1/2\lambda'}\mathcal{D}^{5/2}_{1/2\lambda} = (-1)^{1/2-\lambda'}\mathcal{D}^{5/2}_{-1/2-\lambda'}\mathcal{D}^{5/2}_{1/2\lambda} =$$

21.22

$$= (-1)^{1/2-\lambda'}\sum_{\ell}(\ell 0|\tfrac{5}{2}\ \tfrac{5}{2} - \tfrac{1}{2}\ \tfrac{1}{2})$$

$$(\ell,\lambda-\lambda'|\ \tfrac{5}{2}\ \ \tfrac{5}{2}\ \ -\lambda'\lambda)\mathcal{D}^{\ell}_{0,\lambda-\lambda'}(\theta_1,0)$$

and

$$\mathcal{D}^{5/2*}_{-1/2\lambda'}\mathcal{D}^{5/2}_{-1/2\lambda} = (-1)^{-1/2-\lambda'}\mathcal{D}^{5/2}_{1/2-\lambda'}\mathcal{D}^{5/2}_{-1/2\lambda} =$$

21.23

$$= (-1)^{-1/2-\lambda'}\sum_{\ell}(\ell 0|5/2\ 5/2\ 1/2\ 1/2)$$

$$(\ell,\lambda-\lambda'|5/2\ 5/2\ -\lambda'\lambda)\mathcal{D}^{\ell}_{0,\lambda-\lambda'}(\theta_1,0)$$

If $-1/2 \to +1/2$ the above expression becomes equal to

$$-(-1)^{1/2-\lambda'}\sum_{\ell}(\ell 0\tfrac{5}{2}\ \tfrac{5}{2}-\tfrac{1}{2}\ \tfrac{1}{2})(-1)^{5-\ell}(\ell,\lambda-\lambda'\tfrac{5}{2}\ \tfrac{5}{2}-\lambda'\lambda)\mathcal{D}^{\ell}_{0,\lambda-\lambda'}(\theta_1,0)$$

To flip μ_1,μ_2 need factor $(-1)^{5-\ell}$ (cf. Rose).

Notice that 21.22 and 21.23 are identical except for the factor $-(-1)^{5-\ell}$ which is +1 for ℓ even and -1 for ℓ odd. Hence we have

$$\rho^{(2)}_{\lambda'\lambda}(\theta_1,0) = a(\lambda)a(\lambda')(-1)^{1/2-\lambda'} \times$$

21.24

$$\times\sum_{\ell=\text{even}}(\ell 0|\tfrac{5}{2}\ \tfrac{5}{2}-\tfrac{1}{2}\ \tfrac{1}{2})(\ell,\lambda-\lambda'|\tfrac{5}{2}\ \tfrac{5}{2}-\lambda'\lambda)\left(\frac{4\pi}{2\ell+1}\right)^{1/2}Y_{\ell}^{\lambda-\lambda'}(\theta_1,0)$$

where we have used $\mathcal{D}^{\ell}_{m0}(\phi,\theta) = \left(\frac{4\pi}{2\ell+1}\right)^{1/2}Y_{\ell}^{m*}(\theta,\phi)$.

After some manipulation, explicit expressions for the density matrix elements are found:

$$\rho^{(2)}_{3/2\ 3/2} = 1/6|a(3/2)|^2\{1 + 2/7\ P_2(\theta_1) - 9/7\ P_4(\theta_1)\} \quad 21.25$$

$$\rho^{(2)}_{1/2\ 1/2} = 1/6|a(1/2)|^2\{1 + 8/7\ P_2(\theta_1) + 6/7P_4(\theta_1)\} \quad 21.26$$

$$\rho^{(2)}_{3/2-1/2} = \rho^{(2)}_{-1/2\ 3/2} = -\sqrt{2}/14\ a(-1/2)a^*(3/2)\{7/6 - 1/6\ P_2(\theta_1) - P_4(\theta_1)\} \quad 21.27$$

↑ (provided the a's are real.)

Combining 21.25, 21.26 and 21.14 we obtain

$$\begin{aligned} I(\theta_2,\theta_1) = {} & |a(1/2)|^2+|a(3/2)|^2+2/7(4|a(1/2)|^2+|a(3/2)|^2)P_2(\theta_1) \\ & +3/7(2|a(1/2)|^2-3|a(3/2)|^2)P_4(\theta_1) + \\ & \{|a(1/2)|^2-|a(3/2)|^2+2/7(4|a(1/2)|^2-|a(3/2)|^2)P_2(\theta_1) \\ & +3/7(2|a(1/2)|^2+3|a(3/2)|^2)P_4(\theta_1)\}P_2(\theta_2) \end{aligned} \quad 21.28$$

which is formula (4) of P.R.L. 17, 841 (1966).

At this point certain dynamical assumptions are made to get to the point of determining J^P of $\Sigma(1760)$. These assumptions will determine the ratio between $a(3/2)$ and $a(1/2)$. Jacob and Wick show how to convert from the helicity representation to an ℓ-s representation. Such a change in basis is valid in a non-relativistic limit. For $\Lambda(1520)$ production this is a fairly reasonable approximation at $\sim$ 930 MeV/c since the $\Lambda(1520)$ and π are produced with relatively small momenta. This change of basis will enable us to make statements about the ratio $a(3/2)|a(1/2)$. The basic relation from Jacob and Wick is (JW, B.5)

$$\langle (jm)\ell s|(jm)\lambda_1\lambda_2\rangle = (2\ell+1/2j+1)^{1/2}(j\lambda|\ell_1 s;0\lambda)$$

$$(s_1\lambda_1-\lambda_2|s_1,s_2;\lambda_1,-\lambda_2) \qquad 21.29$$

where a particle of spin $j(\lambda = \lambda_1-\lambda_2)$ decays into a particle of spin $s_2 = 0$ $(\lambda_2 = 0)$ and $s_1 = 3/2(\lambda_1 = \lambda)$. Then

$$(s,\lambda_1-\lambda_2|s_1,s_2;\lambda_1,-\lambda_2) = (-1)^{\lambda-3/2}(\tfrac{3}{2}\ \lambda|0,\ \tfrac{3}{2};0,\lambda) = (-1)^{\lambda-3/2} \qquad 21.30$$

To apply this result we recall that $a(\lambda)$ is the matrix element of a transition operator U in the helicity frame (see Jackson 3.9, 3.10). Thus

$$a(\lambda) = \langle (jm)\lambda_1\lambda_2|U|jm\rangle = \sum_\ell \langle (jm)\lambda_1\lambda_2|(jm)\ell\,\tfrac{3}{2}\rangle \times$$

$$\times\ \langle (jm)\ell\,\tfrac{3}{2}|U|(jm)\rangle = \qquad 21.31$$

$$= (-1)^{\lambda-3/2}\sum_\ell (2\ell+\tfrac{1}{2}\ j+1)^{1/2}(j\lambda\ |\ell,\tfrac{3}{2};0,\lambda)A_\ell$$

for $\eta_{1760} = \eta_{1520}\ \eta_\pi(-1)^\ell$.

The sum over ℓ in 21.31 is constrained by parity conservation in addition to angular momentum conservation. Hence for $\eta_{1760} = -1$.

$$a(\lambda) = (-1)^{\lambda-3/2}\frac{1}{(2j+1)^{1/2}}\sum_{\substack{\ell=\text{odd}\\ =j-3/2,j+1/2\\ =1,3}}(2\ell+1)^{1/2}(j\lambda|\ell,\tfrac{3}{2};0,\lambda)A_\ell \qquad 21.32$$

and for $\eta_{1760} = +1$

$$a(\lambda) = (-1)^{\lambda-3/2}\frac{1}{(2j+1)^{1/2}}\sum_{\substack{\ell=\text{even}\\ =j-1/2,j+3/2\\ =2,4}}(2\ell+1)^{1/2}(j\lambda|\ell,\tfrac{3}{2},0,\lambda)A_\ell \qquad 21.33$$

We now assume that the centrifugal barrier will suppress the contribution from the higher ℓ values, and retain only $\ell = 1$ for $\eta_{1760} = -1$ and $\ell = 2$ for $\eta_{1760} = +1$. We then have

$$\frac{a(3/2)}{a(1/2)} = \begin{cases} \left[\dfrac{2j+3}{3(2j-1)}\right]^{1/2} & \text{for } \eta_{1760}\eta_{1520} = (-1)^{j-1/2} \\ \left[\dfrac{3(2j+3)}{2j-1}\right]^{1/2} & \text{for } \eta_{1760}\eta_{1520} = (-1)^{j+1/2} \end{cases} \qquad 21.34$$

In conclusion

$$\frac{a(3/2)}{a(1/2)} = \begin{cases} \sqrt{2/3} & \text{for } \eta(1760) = -1 \\ \sqrt{6} & \text{for } \eta(1760) = +1 \end{cases} \qquad 21.35$$

Equations 7 and 9 of Fenster et al. follow from 21.28 and 21.34, with the normalization $|a(1/2)|^2 + |a(3/2)|^2 = 1$, 21.36. Similarly for equations 10 and 11 of Fenster et al. In summary

$$\left.\begin{aligned} I(\theta_2,\theta_1) &= 1+\tfrac{4}{5}P_2(\theta_1)+\left[\tfrac{1}{5}+\tfrac{4}{7}P_2(\theta_1)+\tfrac{36}{35}P_4(\theta_1)\right]P_2(\theta_2) \\ I(\theta_3,\theta_1) &= 1+\tfrac{4}{5}P_2(\theta_1)+\left[-\tfrac{7}{10}-\tfrac{1}{5}P_2(\theta_1)\right]P_2(\theta_3) \end{aligned}\right\} \text{ for } J^P = \tfrac{5}{2}^- \qquad 21.37$$

$$\left.\begin{aligned} I(\theta_2,\theta_1) &= 1+\tfrac{20}{49}P_2(\theta_1)-\tfrac{48}{49}P_4(\theta_1)+\left[-\tfrac{5}{7}-\tfrac{4}{49}P_2(\theta_1)+\tfrac{60}{49}P_4(\theta_1)\right]P_2(\theta_2) \\ I(\theta_3,\theta_1) &= 1+\tfrac{20}{49}P_2(\theta_1)-\tfrac{48}{49}P_4(\theta_1)+\left[\tfrac{33}{42}-\tfrac{1}{49}P_2(\theta_1)-\tfrac{48}{49}P_4(\theta_1)\right]P_2(\theta_3) \end{aligned}\right\}$$

$$\text{for } J^P = \tfrac{5}{2}^+. \qquad 21.38$$

These joint correlation functions have been compared with the experimental distributions in the paper by Fenster et al. giving a clear preference for $J^P = \frac{5}{2}^-$ for $\Sigma(1760)$.

References

Original papers describing the discovery of the sequential decay of $\Sigma(1760)$ are, among others,

R. Armenteros, M. Ferro-Luzzi, D.W.G. Leith, R. Levi Setti, A. Minten, R.D. Tripp, H. Filthuth, V. Hepp, E. Kluge, H. Schneider, R. Barloutaud, P. Granet, J. Meyer and J.P. Porte, Phys. Letters 19, 338(1965).

R.B. Bell, R.W. Birge, Y.L. Pan, R.T. Pu, Phys. Rev. Letters 16, 203(1966).

S. Fenster, N.M. Gelfand, D. Harmsen, R. Levi Setti, M. Raymund, J. Doede, W. Männer, Phys. Rev. Letters 17, 841(1966).

More recently, see

W.A. Barletta, Nuclear Physics B40, 45(1972).

PART II

INTRODUCTION TO HIGH ENERGY PHENOMENOLOGY

A. PARTICLE EXCHANGE PROCESSES

22

MANDELSTAM VARIABLES

Consider the scattering process

$$a + b \to c + d \quad \text{or} \quad 1 + 2 \to 3 + 4 \qquad 22.1$$

where the particles have associated <u>four-vectors</u> p_1, p_2, p_3, p_4. It is customary to describe Lorentz-invariant scattering amplitudes in terms of two independent invariant parameters out of the invariants which can be constructed from the four-vectors p_1, p_2, p_3, p_4. Following Collins and Squires, we label all four particles as <u>ingoing</u>, remembering however that any of the actual processes will involve two outgoing anti-particles, relative to such diagram.

There are six possible processes:

$$1 + 2 \to \bar{3} + \bar{4}; \quad 3 + 4 \to \bar{1} + \bar{2} \qquad 22.2a$$

$$1 + 3 \to \bar{2} + \bar{4}; \quad 2 + 4 \to \bar{1} + \bar{3} \qquad 22.2b$$

$$1 + 4 \to \bar{2} + \bar{3}; \quad 2 + 3 \to \bar{1} + \bar{4} \qquad 22.2c$$

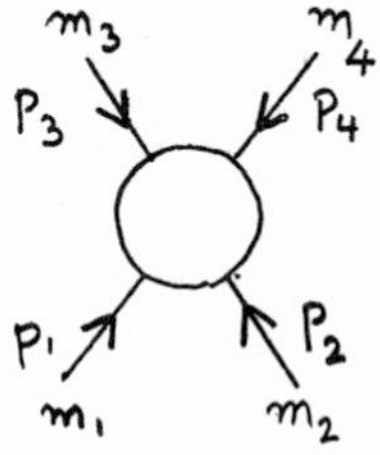

Define the invariants:

$$s = (p_1+p_2)^2 = (p_3+p_4)^2 \qquad 22.3a$$

$$t = (p_1+p_3)^2 = (p_2+p_4)^2 \qquad 22.3b$$

$$u = (p_1+p_4)^2 = \underbrace{(p_2+p_3)^2} \qquad 22.3c$$

↑ from energy-momentum conservation.

In the c.m.s. of particle 1 and 2 $\rightarrow p_1 = (E_1, \vec{q}_{s12}),\quad p_2 = (E_2, -\vec{q}_{s12})$

Similarly in the c.m.s. of 3 and 4 $\rightarrow p_3 = (E_3, \vec{q}_{s34}),\quad p_4 = (E_4, -\vec{q}_{s34})$ 22.4

Since for 4-vectors a and b

$a \cdot b = a_\mu b_\mu = a_4 b_4 - \vec{a}\cdot\vec{b}$ 22.5

$p_1^2 = E_1^2 - q_{s12}^2 = m_1^2$ 22.6a

$p_2^2 = E_2^2 - q_{s12}^2 = m_2^2$ 22.6b

$p_3^2 = E_3^2 - q_{s34}^2 = m_3^2$ 22.6c

$p_4^2 = E_4^2 - q_{s34}^2 = m_4^2$ 22.6d

$s = (E_1 + E_2)^2$

$= (\text{c.m. energy})^2$ 22.7

"<u>s-channel</u>."

From 22.3a: $s = p_1^2 + p_2^2 + 2p_1 \cdot p_2 = m_1^2 + m_2^2 + 2p_i \cdot p_2$ (a)

From 22.3b: $t = m_1^2 + m_3^2 + 2p_1 \cdot p_3$ (b) 22.8

From 22.3c: $u = m_1^2 + m_4^2 + 2p_i \cdot p_4$ (c)

$s+t+u = m_1^2 + m_2^2 + m_3^2 + m_4^2 + 2m_1^2 + 2p_1(p_2 + p_3 + p_4)$ 22.9

But because of four-momentum conservation

$p_2 + p_3 + p_4 = -p_1$ 22.10 and from 22.6a

$s + t + u = \sum_i m_i^2 = \Sigma$ 22.11

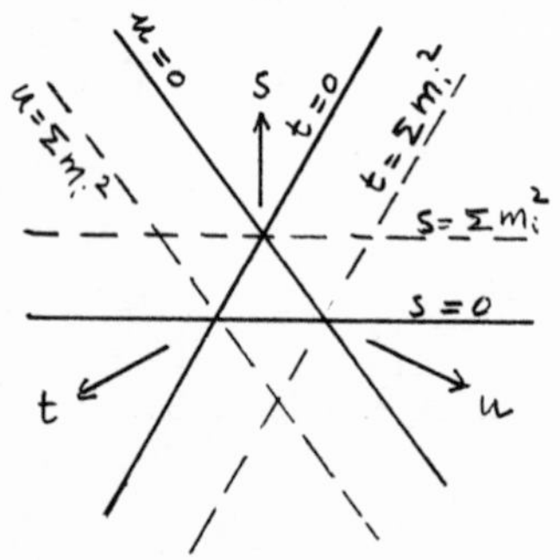

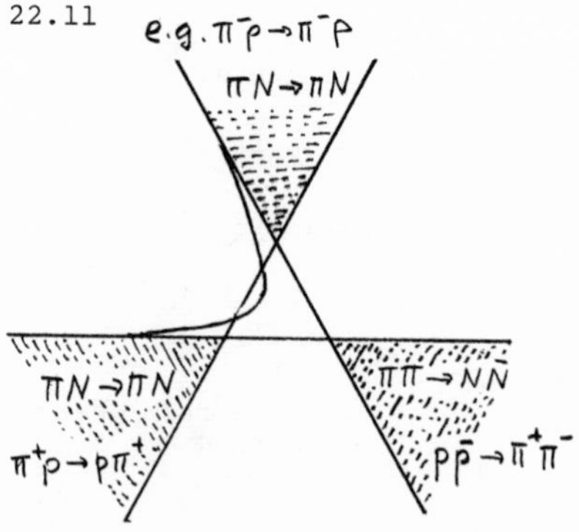

Useful relations which follow from above:

$$p_1 \cdot p_2 = 1/2\,(s-m_1^2-m_2^2) \tag{22.12}$$

$$m_1^2+p_1\cdot p_2 = p_1\cdot(p_1+p_2) = E_1\sqrt{s} \tag{22.13}$$

$$E_1 = \frac{1}{2\sqrt{s}}(s+m_1^2-m_2^2) \tag{22.14}$$

and similar ones for ⟶ $E_2,\ E_3,\ E_4$

with mass indices ⟶ 21 34 43

From 22.14 and 22.6a:

$$q_{s12}^2 = 1/4s\left[s-(m_1+m_2)^2\right]\left[s-(m_1-m_2)^2\right] \tag{22.15}$$

$$q_{s34}^2 = 1/4s\left[s-(m_3+m_4)^2\right]\left[s-(m_3-m_4)^2\right] \tag{22.16}$$

Note that for $s \gg m_1, m_2$, $q \approx 1/2\sqrt{s}$ 22.17

From $t = m_1^2+m_3^2+2p_1\cdot p_3$, follows

$$t = m_1^2+m_3^2+2E_1E_3-2\vec{q}_{s12}\cdot\vec{q}_{s34} = m_1^2+m_3^2+2E_1E_3-2q_{s12}q_{s34}\cos\theta_s \tag{22.18}$$

where θ_s is the c.m. scattering angle between particles 1 and 3. Using 22.14 one finds

$$\cos\theta_s = \frac{s^2+s(2t-\Sigma)+(m_1^2-m_2^2)(m_3^2-m_4^2)}{4s\ q_{s12}q_{s34}} \tag{22.19}$$

If all masses are equal, 22.16 yields

$$q_s^2 = \frac{s-4m^2}{4} \tag{22.20}$$

and

$$\cos\theta_s = 1 + \frac{t}{2q_s^2} = 1 + \frac{2t}{s-4m^2} \tag{22.21}$$

or

$$t = 2q_s^2(1-\cos\theta_s) \qquad 22.22$$

which is the well-known relation for the square of the momentum transfer. This approach outlines a simple way to obtain all quantities of interest.

Physical regions: They are indicated by the dashed areas in the Mandelstam plot on p. 232 for, e.g., $\pi N \to \pi N$.

For the s-channel: $q_{s12}^2 > 0 \to s > (m_1+m_2)^2$ 22.23;

from 22.15, $-1 \leq \cos\theta_s \leq 1$ 22.24.

For equal masses this implies

$$-(s-4m^2) \leq t \leq 0 \quad \text{or} \quad -4q_s^2 \leq t \leq 0 \qquad 22.25$$

Note that since in this case $s + t + u = 4m^2$, the line $-t = -(s - 4m^2)$ corresponds to $u = 0$.

The kinematic quantities derived so far have been expressed in terms of the s-channel variables 22.4. In summary, for the reaction $1 + 2 \to \bar{3} + \bar{4}$

$$(p_1 \quad p_2 \quad p_3 \quad p_4)$$

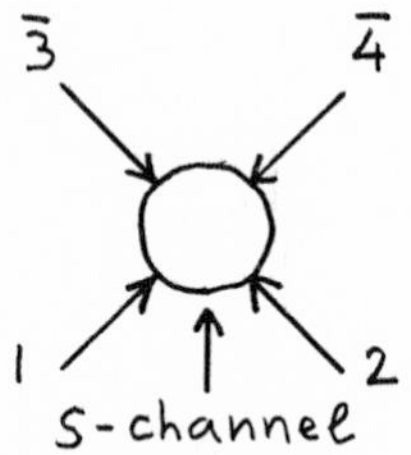

$$s = (p_1+p_2)^2 = (\text{c.m. energy})^2$$
$$t = (p_1+p_3) = (\text{momentum transfer})^2$$
$$u = (p_1+p_4) = (\text{momentum transfer})^2$$
22.26

Implicit in the set of relations (2) and (3) is however the additional concept of crossing. Note that if p_a is the 4-momentum of particle a or $\bar{a}$, $-\bar{p}_a$ is the 4-momentum of antiparticle $\bar{a}$ or a, where

$$p_a = (E_a, \vec{q}_a) \qquad \begin{matrix} a, & \bar{a} \\ \downarrow & \downarrow \\ \bar{a}, & a \end{matrix} \qquad 22.27$$
$$-\bar{p}_a = (-E_a, -\vec{q}_a)$$

Reactions 22.2b, 22.2c:

$1 + 3 \to \bar{2} + \bar{4} \to$ t-channel	define the <u>crossed</u> channels
$1 + 4 \to \bar{2} + \bar{3} \to$ u-channel	t and u respectively.

It is important to obtain relations similar to those derived above, in terms of quantities defined in the respective channels. The meaning of s,t,u is however interchanged as follows:

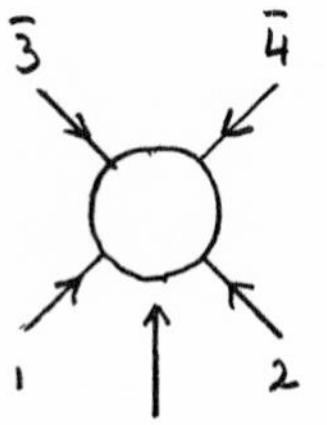

s-channel

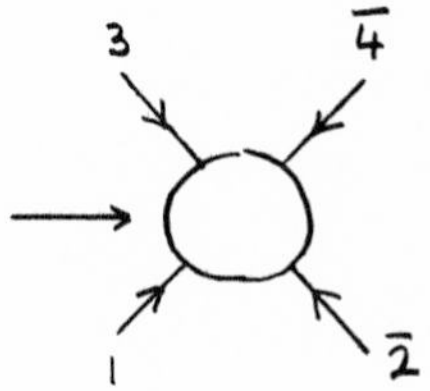

t-channel

$$t = (p_1+p_3)^2 \quad \longleftrightarrow \quad s = (p_1-\bar{p}_2)^2 = (p_1+p_2)^2 = \begin{pmatrix}\text{mom.} \\ \text{transfer}\end{pmatrix}^2$$

$$\overbrace{1 + 2} \to \bar{3} + \bar{4} \qquad \overbrace{1 + 3} \to \bar{2} + \bar{4} \qquad 22.28$$

$$s = (p_1+p_2)^2 \quad \longleftrightarrow \quad t = (p_1-\bar{p}_3)^2 = (p_1+p_3)^2 = \begin{pmatrix}\text{c.m.} \\ \text{energy}\end{pmatrix}^2$$

while u remains unchanged.

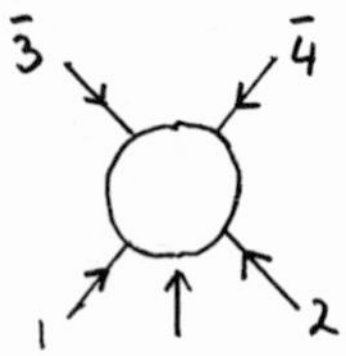

s-channel

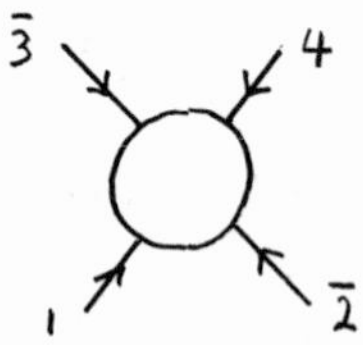

u-channel

$u = (p_1+p_4)^2 \quad \longleftrightarrow \quad s = (p_1-\bar{p}_2)^2 = (p_1+p_2)^2 = \begin{pmatrix}\text{mom.}\\ \text{transfer}\end{pmatrix}^2$

$1 + 2 \rightarrow \bar{3} + 4 \qquad\qquad 1 + 4 \rightarrow \bar{3} + \bar{2}$

$s = (p_1+p_2)^2 \quad \longleftrightarrow \quad u = (p_1-\bar{p}_4)^2 = (p_1+p_4)^2 = \begin{pmatrix}\text{c.m.}\\ \text{energy}\end{pmatrix}^2$

while t remains unchanged.

Being careful about the sign of the components of the 4-momenta involved, by a procedure entirely identical to the previous one, we obtain, among others, the cosines of the scattering angles θ_t and θ_u:

$$\cos\theta_t = \frac{t^2+t(2s-\Sigma)+(m_3-m_1)(m_4-m_2)}{4t\, q_{t13}\, q_{t24}} \qquad 22.30$$

$$\cos\theta_u = \frac{u^2+u(2t-\Sigma)+(m_1-m_4)(m_2-m_3)}{4u\, q_{u14}\, q_{u23}} \qquad 22.31$$

Crossing symmetry, particle exchange.

Define Lorentz invariant scattering amplitude, and express cross sections in terms of s,t,u.

Non-relativistic	Lorentz invariant	
$f(k,\theta)$	$T(s,t) = 8\pi\sqrt{s}\ f(k,\theta)$	22.32

Differential cross section as a function of t:

$$\frac{d\sigma}{dt} = \frac{d\sigma}{d\Omega}\frac{d\Omega}{dt} \qquad 22.33$$

$$t = -2k^2(\cos\theta)$$
$$dt = 2k^2 d\cos\theta \qquad 22.34$$

$$\frac{d\sigma}{dt} = \frac{d\sigma}{d\Omega}\frac{2\pi\ d\cos\theta}{2k^2 d\cos\theta} = \frac{\pi}{k^2}\frac{d\sigma}{d\Omega} \qquad 22.35$$

The optical theorem, in terms of the t-variable:

$$\sigma_{tot} = \frac{4\pi}{k}\,\mathrm{Im}\, f(\underset{\theta=0}{\underset{\uparrow}{k,0}}) \to 4\sqrt{\pi}\,\mathrm{Im}\, f(\underset{t=0}{\underset{\uparrow}{k,0}}) \qquad 22.36$$

In Lorentz-invariant form:

$$\sigma_{tot} = \frac{4\pi}{k}\,\mathrm{Im}\, f(k,0) = \frac{4\pi}{k}\frac{1}{8\pi\sqrt{s}}\,\mathrm{Im}\, T(s,0) = \frac{1}{2k\sqrt{s}}\,\mathrm{Im}\, T(s,0) \qquad 22.37$$

Differential cross sections:

$$\left.\frac{d\sigma}{d\Omega}\right]_{\theta=0} = |\mathrm{Re}\ f(0)|^2 + |\mathrm{Im}\ f(0)|^2 = |\mathrm{Re}\ f(0)|^2 + \frac{\sigma^2 k^2}{(4\pi)^2} \qquad 22.38$$

$$\left.\frac{d\sigma}{dt}\right]_{t=0} = \frac{\pi}{k^2}|\mathrm{Re}\ f(0)|^2 + \frac{\sigma^2}{16\pi} \qquad 22.39$$

Lorentz invariant normalization:

$$\frac{d\sigma}{dt} = \frac{\pi}{k^2}\ |f(\theta)|^2 \to \frac{\pi}{k^2}\frac{1}{64\pi^2 s}|T(s,t)|^2 = \frac{1}{64\pi k^2 s}|T(s,t)|^2 \qquad 22.40$$

Consider now for example the πN reactions:

$$\left.\begin{array}{l}\pi^- p \to \pi^0 n \\ 1+2 \to \bar{3}+\bar{4}\end{array}\right\}$$ <u>s-channel</u> with our previous conventions. 22.41

Call $T(s,t)$, $T(p_1,p_2,p_3,p_4)$ } the s-channel Lorentz-invariant scattering amplitude.

For the crossed channel t,

$$\left.\begin{array}{l} \pi^-\pi^0 \rightarrow \bar{p}n \\ 1+3 \rightarrow \bar{2}+\bar{4} \end{array}\right\} \text{ t-channel} \qquad 22.42$$

Call $\bar{T}(\bar{s},\bar{t})$, $\bar{T}(p_1,-\bar{p}_2,-\bar{p}_3,p_4)$ } the t-channel Lorentz invariant scattering amplitude. 22.43

The principle of crossing-symmetry implies that the s- and t-channel are connected, in fact by the same scattering amplitude:

$$T(s,t) = \bar{T}(\bar{s},\bar{t}) \qquad 22.44$$

This means that knowing the analytic form of $T(s,t)$, the same analytic form will describe the crossed-channel reaction, provided that the signs of the four-momenta of the particles which become antiparticles in the crossing are properly inverted. Since the meaning of s,t is reversed in going from $s \rightarrow t$, as shown at page 235, the values of the variables in T and $\bar{T}$ are different, each one being confined to the physical regions of the respective channels. Alternatively, since the physical regions of the s- and t-channel do not overlap, 22.44 says that $T(s,t)$ can be analytically continued from the physical region of one channel to the other. 22.44 is a result of quantum field theory (see, e.g., Van Hove, CERN Report 68-31, 25 July 1968).

Crossing symmetry plays a fundamental role in explaining a large class of phenomena occurring at high energy.

The relevant feature is that singularities of the scattering amplitude, such as poles, in one channel, may enhance the amplitude in the physical regions near the singularities, in the crossed channels.

As an example, will consider the one meson exchange model, without entering into great detail for this purpose. Start with the crossed channel reaction:

$\pi^- \pi^0 \rightarrow \bar{p}n$ (22.42)

The partial wave expansion for reaction 22.42 is of the form:

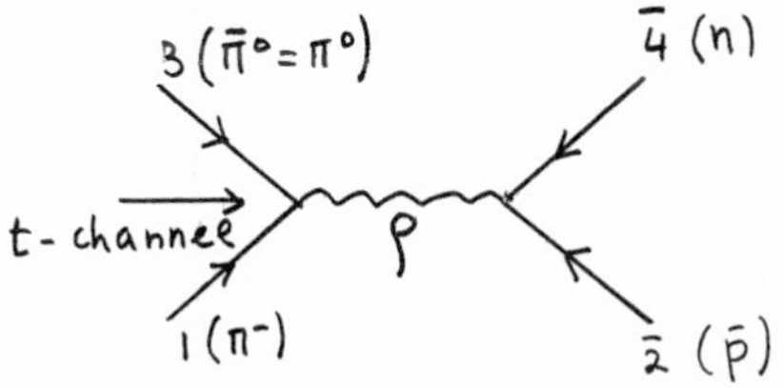

$$\bar{T}(\bar{s},\bar{t}) = 8\pi\sqrt{\bar{s}}\,\frac{1}{\sqrt{k_t k_t'}}\sum_{\ell=0}^{\infty}(2\ell+1)\bar{f}(\ell,E_t)P_\ell(\cos\theta_t) \qquad 22.45$$

where k_t, k_t' are the entrance and exit channel momenta. Note $\cos\theta_t$, t-channel variable.

At energy E_t near the ρ-meson mass m_ρ (this is an unphysical energy, below the threshold 2M, M = nucleon mass), the scattering amplitude is expected to be dominated by the $\ell = 1$ partial wave in which the ρ meson occurs as a resonance. The one-ρ-meson exchange model for the s-channel reaction 22.41: $\pi^- p \rightarrow \pi^0 n$ consists in assuming that the t-channel amplitude is dominated by this resonating p-wave, and to invoking crossing symmetry to obtain the amplitude in the s-channel. More specifically

$$\bar{T}(\bar{s},\bar{t}) \approx \bar{T}_\rho(\bar{s},\bar{t}) = K_\rho(\bar{s})P_{\ell=1}(\cos\theta_t) \qquad 22.46$$

where

$$K_\rho(\bar{s}) = 8\pi\sqrt{s}\,\frac{1}{\sqrt{k_t k_{t'}}}\cdot 3\cdot\bar{f}_\rho(\ell=1,E_t) \qquad 22.47$$

and the partial wave amplitude $\bar{f}_\rho$ is described, for example, by the Born term for the exchange of a vector meson, in the lowest order Feynman diagram:

$$\bar{f}_\rho = B^\mu_{\pi^-\rho\pi^0}(t,m_{\pi^-},m_{\pi^0})\frac{g_{\mu\nu}-p_\mu p_\nu/m_\rho^2}{t-m_\rho^2}B^\nu_{\bar{p}\rho n}(t,m_{\bar{p}},m_n) \qquad 22.48$$

for $(\bar{s} = t)$. B^μ, B^ν are the vertex functions, 4-vectors in this case. Note the propagator $\frac{1}{t-m_\rho^2}$ which gives an enhancement in the amplitude at $t \sim m^2$. Note also that 22.48 is independent of $\cos\theta_t$.

Now, making use of the crossing relation 22.44, the amplitude 22.46 is assumed to be a good approximation to the scattering amplitude for the s-channel reaction $\pi^- p \to \pi^0 n$. $K_\rho(\bar{s}) \to K_\rho(t)$. For $\cos\theta_t$ we obtain from 22.30, for $m_1 = m_3 = \mu$, $m_2 = m_4 = M$:

$$\cos\theta_t = \frac{s+\frac{t}{2}-\mu^2-M^2}{\frac{1}{2}\sqrt{(t-4\mu^2)}\,\sqrt{(t-4M^2)}} \qquad 22.49$$

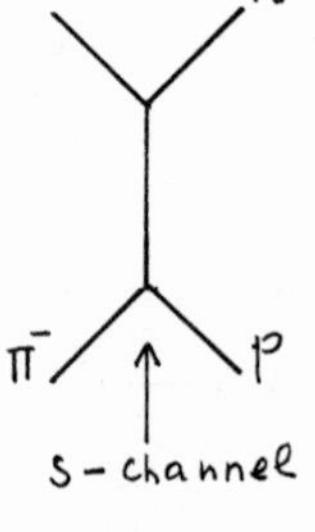

For the s-channel reaction then

$$T(s,t) = K_\rho(t)P_{\ell=1}(\cos\theta_t) = K_\rho(t)\frac{-2(s+\frac{t}{2}-\mu^2-M^2)}{\sqrt{4\mu^2-t}\,\sqrt{4M^2-t}} \qquad 22.50$$

where a factor i has been extracted from each square root in the denominator, to obtain roots of positive numbers.

Through $K_\rho(t)$, the s-channel amplitude still contains the propagator $\frac{1}{t-m_\rho^2}$. For t near m_ρ^2, $K_\rho(t)$ becomes very large. For t fairly close to the ρ pole, the approximation of keeping only the ρ-exchange contribution may be adequate. This becomes more and more dubious for t very different from m^2. To the extent that the approximation is valid also for $t < 0$ (the physical region for the s-channel), the propagator enhances the s-channel differential cross section at small $|t|$, namely predicts a peak in $\frac{d\sigma}{dt}$ in the forward direction. This is a well known feature of the experimental data. The above considerations are at the base of the so-called "peripheral exchange" or "peripheral model," used in the interpretation of a variety of processes of production of resonances, namely inelastic processes, in addition to elastic scattering.

The particular model discussed above is beset with several shortcomings however; first the t-dependence of $d\sigma/dt$ is not as sharp as indicated experimentally; second the s-dependence of the cross section, for fixed t, is wrong. In fact the amplitude 22.50, for fixed t, behaves as a constant s. Since from 22.40

$$\frac{d\sigma}{dt} = \frac{1}{64\pi\, k^2 s}\,|T(s,t)|^2 \xrightarrow[\text{large } s]{} \frac{1}{16\pi s^2}\,|T(s,t)|^2 \qquad 22.51$$

The prediction of 22.50 is

$$\left.\frac{d\sigma}{dt}\right|_{\text{fixed } t} \xrightarrow[s \text{ large}]{} \text{constant}$$

Experimentally instead

$$\frac{d\sigma}{dt} \sim s^{-1}.$$

While the bad features of the t-dependence in the OME model can be cured (absorption model, etc.), the trouble with the energy dependence can only be overcome by recourse to, e.g., the Regge pole model. The essential feature however to be retained for future reference is the notion that the particle exchange, or pole-exchange in the t-channel (for the meson exchange case) determines the behaviour of the s-channel amplitude.

The s-channel behaviour will of course be determined also by the presence of direct channel resonances, as seen in detail in Part I.

The effect of poles in the various channels can be summarized as follows: (consider meson-baryon scattering)

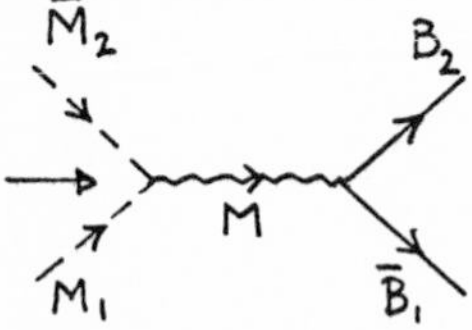

Meson exchange
→ "forward peak"

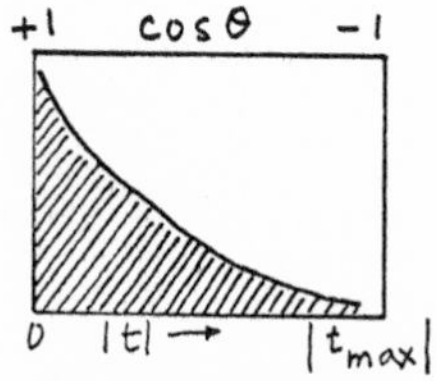

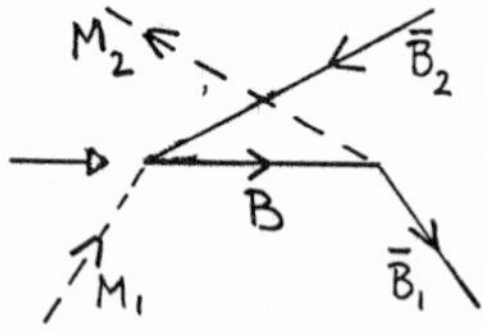

Baryon exchange
→ "backward peak"

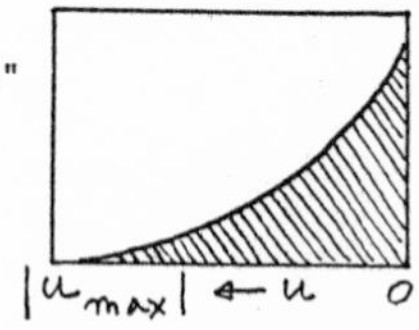

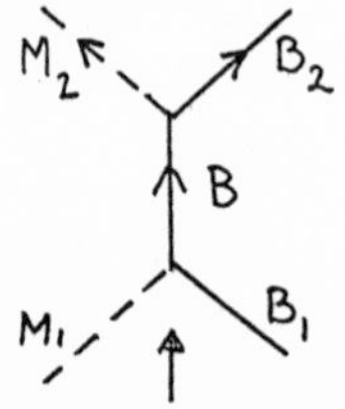

Direct channel
resonance

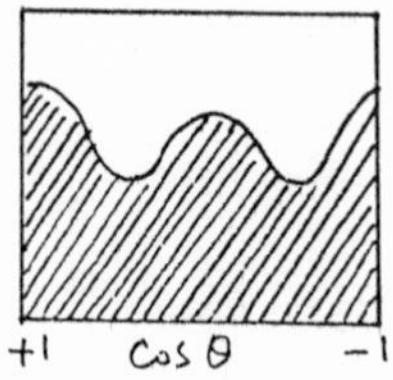

The remarkable feature of this predicted correspondence between singularities of the s-matrix and experimental effects in the form of forward or backward peaks is that it is verified for essentially all explored processes. In reverse, the absence of forward or backward peaks that would result from the exchange of a particular set of quantum numbers, should on the same grounds reflect the absence in nature of physical particles with those quantum numbers. This is indeed the case.

The suppression caused by the absence of particles to be exchanged is typically two orders of magnitude or more. For example:

Reaction	t-channel exchange	u-channel exchange
$\pi^- p \rightarrow \pi^- p$; $\rho^- p$	known mesons $s = 0$	Δ
$K^+ p \rightarrow K^+ p$; $K^{*+} p$	known mesons $s = 0$	known hyperons $s = -1$
$\pi^- p \rightarrow K^0 \Lambda$	known mesons $s = +1$	known hyperons $s = -1$
Peak:	forward observed	backward observed
$K^- p \rightarrow K^- p$; $K^{*-} p$	known mesons $s = 0$	No known baryons $s = +1$
$\bar{p}p \rightarrow \bar{\Lambda}\Lambda$	known mesons $s = +1$	No known baryons $B = 2,\ s = -1$
Peak:	forward observed	backward absent

A few examples of the experimental evidence for the above follow:

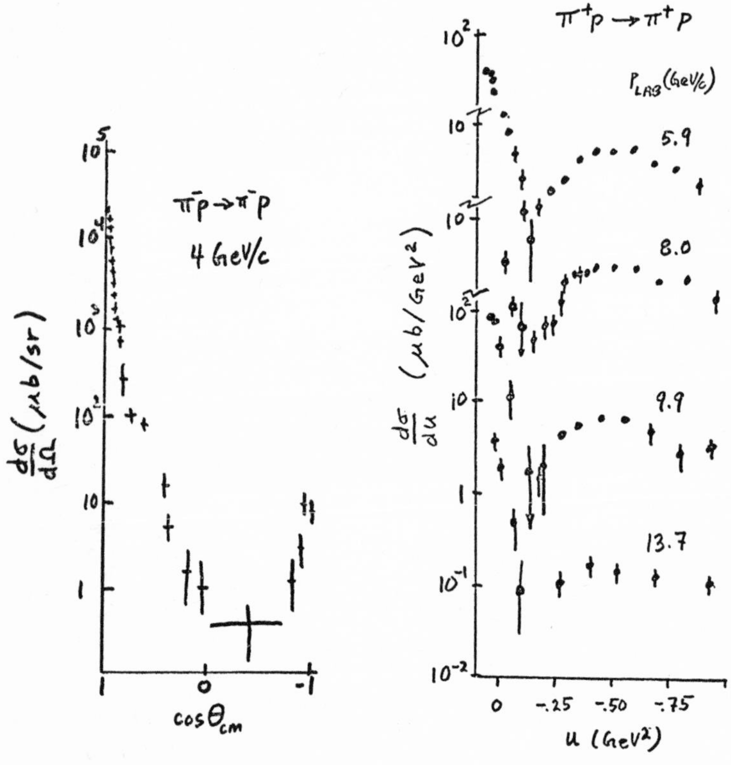

Example of forward and backward peaks in $\pi^- p \to \pi^- p$ and of backward peaks in $\pi^+ p \to \pi^+ p$, adapted from a compilation of data from the literature by V.D. Barger and D.B. Cline, 1969.

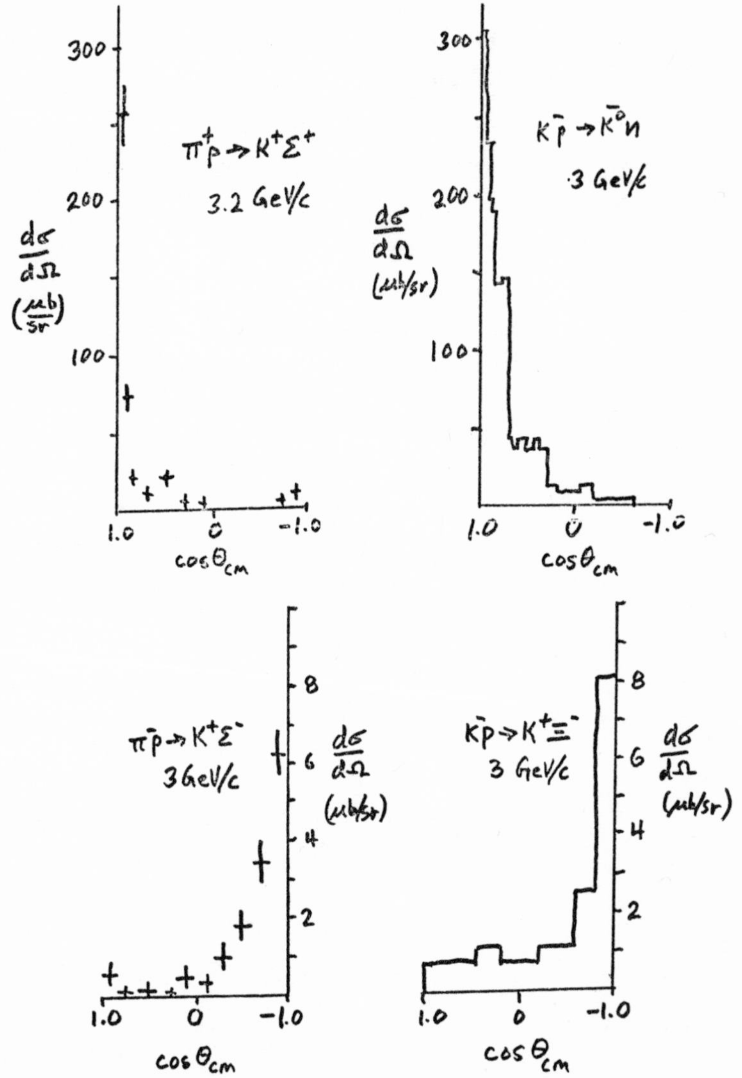

Illustration of the suppression of forward or backward peaks attributable to "forbidden" particle exchange. Adapted from an illustration of data from the literature by V. Barger, Rev. Mod. Physics 40, 129(1968).

References

P.D.B. Collins and E.J. Squires. Regge Poles in Particle Physics. Springer-Verlag, Berlin-Heidelberg-New York, 1968.

H. Pilkuhn. The Interactions of Hadrons. North-Holland Publishing Co., Amsterdam. John Wiley & Sons, Inc., New York, 1967.

Of particular value, in the preparation of this lecture, have been the lectures by

B.E.Y. Svensson, High Energy Phenomenology and Regge Poles, Proceedings of the 1967 CERN School of Physics at Rättvik. CERN Report No. 67-24, Vol. II.

See also

V.D. Barger and D.B. Cline. Phenomenological Theories of High Energy Scattering. An Experimental Evaluation. W.A. Benjamin, Inc., New York, 1969.

23

REGGE POLE PHENOMENOLOGY

Generalities. The Regge pole model connects two apparently unrelated classes of phenomena:

a) Particle classification

b) High energy scattering and exchange processes

a) Particles (bound states, resonances) having the same internal quantum numbers (B,I,G-parity,S,P) are correlated by Regge trajectories, where they appear with Spin J differing by 2 units ($\Delta J = 2$).

b) Quasi-two-body reactions are predicted to be dominated by the exchange of one or more of these trajectories (in the same sense as the one-meson exchange model discussed previously).

The rotational rule $\Delta J = 2$ involves an additional quantum number τ called signature

$$\tau = (-)^{J-1/2} \text{ for baryons,} \quad \tau = (-)^{J} \text{ for mesons} \qquad 23.1$$

$$J = 1/2,\ 5/2,\ 9/2,\ \ldots \qquad J = 0,\ 2,\ 4,\ \ldots$$

The Chew-Frautschi plot is a plot of J versus $(\text{mass})^2$. The Regge-pole model states that there exist a complex function $\alpha(M_R)$ such that

$$\text{Re}\ \alpha(M_R) = J_R \qquad 23.2$$

The functional dependence is deduced empirically. The trajectory is presumed to be an analytic function of $t, \alpha(t)$ for <u>mesons</u>, and $\alpha\sqrt{u}$ for <u>baryons</u>.

For baryons $(u = m^2)$

$\mathrm{Re}\,\alpha(\sqrt{u}) \simeq a+bu$ 23.3

For mesons

$\mathrm{Re}\,\alpha(t) \simeq a+bt$ 23.4

(See attached plots.)

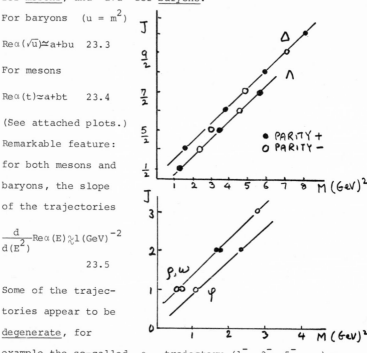

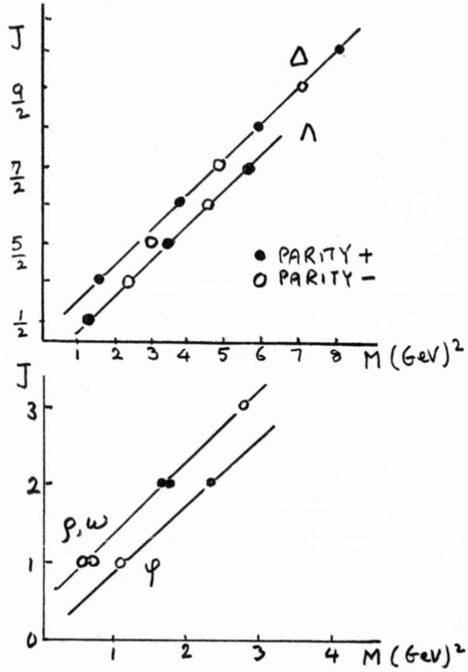

Remarkable feature: for both mesons and baryons, the slope of the trajectories

$$\frac{d}{d(E^2)}\mathrm{Re}\,\alpha(E) \approx 1\,(\mathrm{GeV})^{-2}$$ 23.5

Some of the trajectories appear to be <u>degenerate</u>, for example the so-called ρ_V trajectory $(1^-, 3^-, 5^-, \ldots)$ almost coincides with the π_T trajectory $(2^+, 4^+, \ldots)(A_2)$. Similarly the so-called $\Lambda_\alpha(1/2^+, 5/2^+, \ldots)$ and the $\Lambda_\gamma(3/2^-, 7/2^-, \ldots)$. The first state on the π_T or A_2 trajectory has $J = 2^+$. An extrapolation to 0^+ would place such an hypothetical particle at negative M^2. Such a state is called a <u>ghost state</u>.

<u>Regge Poles in Potential Scattering</u>

Regge has considered the solutions of the radial Schroedinger equation for arbitrary complex ℓ-values, for

Yukawa shape potential $V_Y(r) - g\frac{1}{r}e^{-\nu r}$ or superposition thereof. Relevant results are:

a) The radial Schroedinger equation

$$-\frac{1}{2\mu}\left[\frac{d^2}{dr^2} - \frac{\ell(\ell+1)}{r^2}\right]u_\ell(r) + V(r)u_\ell(v) = Eu_\ell(r) \qquad 23.6$$

has solutions for arbitrary complex ℓ, provided Re$\ell \geq -1/2$. Thus a partial wave amplitude $f(\ell,E)$ can be defined in the complex ℓ-plane.

b) $f(\ell,E)$ is a meromorphic function of ℓ (analytic except for isolated poles) for Re$\ell > -1/2$, continuous for Re$\ell = -1/2$.

c) The poles of $f(\ell,E)$(Regge poles) occur on the real ℓ-axis if E is below the scattering threshold E_{th} ($E < 0$ non-relativistically). The residue at the pole is real. Also, poles occur in the upper half plane, Im$\ell > 0$, if E is above E_{th}($E > 0$). The number of poles in the upper half plane is finite.

d) As $|\ell| \to \infty$, $|f(\ell,E)|$ is well behaved.

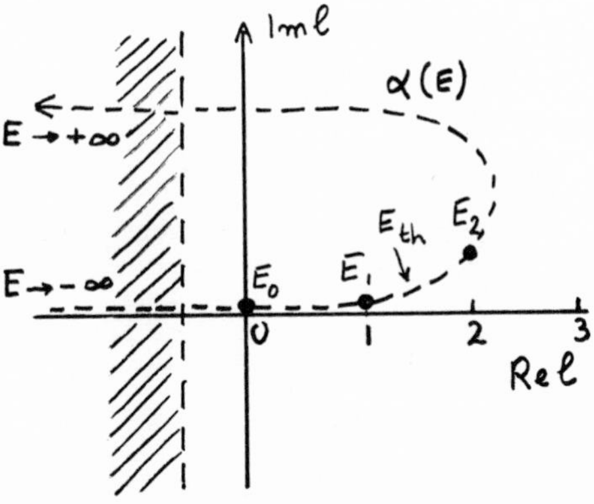

Motion of one Regge pole in the complex ℓ-plane as a function of E.

The contribution given by Regge poles to the scattering amplitude can be evaluated by means of the <u>Sommerfeld-Watson</u> transform. This transforms the sum over partial waves at integral angular-momenta into an integral along a path C in the complex ℓ-plane. Consider the usual partial wave expansion

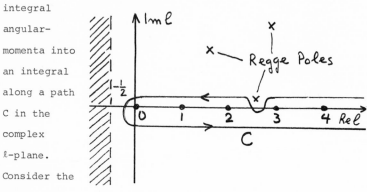

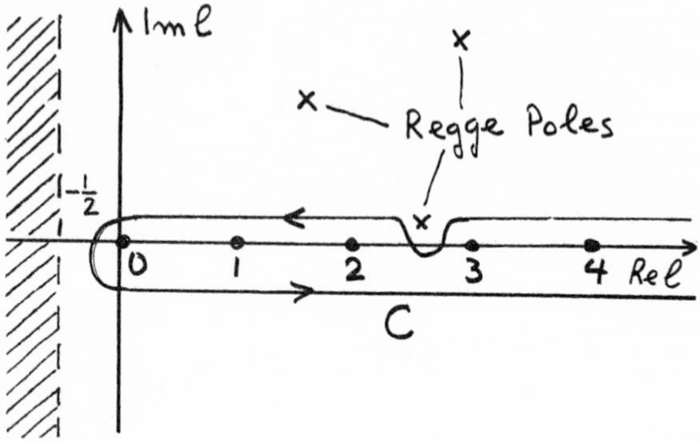

$$f(\theta,E) = \sum_{\ell=0}^{\infty} (2\ell+1)F(\ell,E)P_\ell(\cos\theta) \qquad 23.7$$

For ℓ = complex

$$f(\cos\theta, E) = \frac{i}{2}\int_C \frac{(2\ell+1)F(\ell,E)P_\ell(-\cos\theta)}{\sin\pi\ell}\,d\ell \qquad 23.8$$

This can be seen by use of Cauchy's integral theorem:

$$\oint_C = 2\pi i \,\Sigma \text{ residues } (\ell = n) \qquad 23.9$$

Note that the contour C encircles the poles of $\sin\pi\ell$, but not the Regge poles. In fact consider the integrand in 23.8 around an integer value n of ℓ:

$$\ell \approx n + x,\ x \ll 1; \quad \sin\pi\ell = \sin\pi(n + x) =$$

$$\cos\pi n \sin\pi x + \sin\pi n \cos\pi x = (-1)^n x \qquad 23.10$$

$$g(\ell) = \frac{i}{2}\,\frac{(2\ell+1)F(\ell,E)P_\ell(-\cos\theta)}{\sin\pi\ell} \to \frac{i}{2}\,\frac{(2n+1)F(n,E)(-1)^n P_n(-\cos\theta)}{\pi x} \qquad 23.11$$

But $(-1)^n P_n(-\cos\theta) = P_n(\cos\theta)$. 23.12

The residue of 23.11 is just the coefficient of $1/x$, then, reversing the sense of circulation in the contour integral:

$$\oint_C = -2\pi i \frac{i}{2} \sum \frac{(2n+1)F(n,E)P_n(\cos\theta)}{\pi} = f(\theta,E) \qquad 23.13$$

Now we have to take care of the Regge poles, which have been avoided so far.

The previous contour, closing at $\mathrm{Re}\,\ell \to \infty$, is now deformed to include two arcs of a circle in the first and fourth quadrant. The closure takes place by means of an integral parallel to the $\mathrm{Im}\,\ell$ axis, along $\mathrm{Re}\,\ell = -1/2 + \varepsilon$, $\varepsilon > 0$, and small.

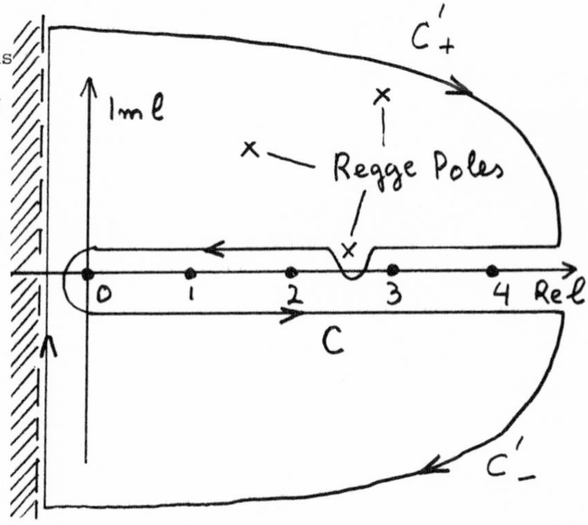

Applying Cauchy's theorem to the new path, we obtain

$$\oint_{C'} = \int_C - \int_{C'+} + \int_{C'-} - \frac{i}{2}\int_{\mathrm{Re}\,=-\frac{1}{2}+\varepsilon} \frac{(2\ell+1)F(\ell,E)P_\ell(-\cos\theta)}{\sin\pi\ell}\,d\ell$$

$$= -2\pi i \Sigma_i \ \mathrm{Res.}\ g(\ell=\alpha_i). \qquad 23.14$$

As the radius of the arcs C'_+ and $C'_- \to \infty$ the integrals along such arcs vanish. We then have from 23.13 and 23.14

$$\oint_C = f(\theta,E) = -\Sigma_i \mathrm{Res}\; g(\ell=\alpha_i) + \frac{i}{2}\int_{-1/2-i\infty}^{+1/2+i\infty} \frac{(2\ell+1)F(\ell,E)P_\ell(-\cos\theta)}{\sin\pi\ell}\, d\ell \qquad 23.15$$

The last term in 23.15 is called background integral, B.I.(cos θ,E). The residues of the Regge poles at $\ell = \alpha_i$, for α_i not integer, will be of the form:

$$\mathrm{Res}\; g(\ell=\alpha_i) = \frac{i}{2}\,\frac{[2\alpha_i(E)+1]R_i(E)P_{\alpha_i(E)}(-\cos\theta)}{\sin\pi\alpha_i(E)} \qquad 23.16$$

or in short

$$\mathrm{Res}\; g(\ell=\alpha_i) = \frac{\beta_i(E)P_{\alpha_i(E)}(-\cos\theta)}{\sin\pi\alpha_i(E)} \qquad 23.17$$

In conclusion:

$$f(\theta,E) = \Sigma_i\,\frac{\beta_i(E)P_{\alpha_i(E)}(-\cos\theta)}{\sin\pi\alpha_i(E)} + \mathrm{B.I.}(\cos\theta,E) \qquad 23.18$$

This relation is usually called the Regge-Sommerfeld-Watson representation of the scattering amplitude. 23.18 is valid for all values of (-cos θ), therefore including the unphysical region. In fact

$$P_\ell(x) \underset{x\to\pm\infty}{\to} (2x)^\ell \frac{\Gamma(\ell+\frac{1}{2})}{\sqrt{\pi}\;\Gamma(\ell+1)}\,, \qquad 23.19$$

$\mathrm{Re}\;\ell > -\frac{1}{2}$. This relation will become relevant later.

We are interested now in showing how a moving Regge pole can give rise to bound states and resonances. Consider the contribution of one pole in the complex ℓ-plane, to the scattering amplitude:

$$f(\cos\theta, E) = \ldots + \frac{\beta(E)}{\sin \pi\alpha(E)} P_{\alpha(E)}(-\cos\theta) + \ldots \qquad 23.20$$

We project now the scattering amplitude onto a particular partial wave $f_\ell(E)$ for ℓ = integer:

$$f_\ell(E) = \int_{-1}^{+1} \frac{\beta(E)}{\sin \pi\, \alpha(E)} P_\ell(\cos\theta) P_{\alpha(E)}(-\cos\theta)\, d\cos\theta$$

$$= \frac{1}{\pi} \frac{\beta(E)}{[\alpha(E)+\ell+1][\alpha(E)-\ell]} \qquad 23.21$$

where we have made use of the integral formula

$$\frac{1}{2}\int_{-1}^{+1} P_\ell(x) P_\alpha(-x)\, dx = \frac{1}{4} \frac{\sin \pi\alpha}{(\alpha-\ell)(\alpha+\ell+1)} \qquad 23.22$$

It is important to note that one Regge pole contributes, in general, to all partial waves, as can be seen from 23.21. Now, when $\alpha(E)$, for a certain energy E_R, approaches the integer ℓ, we can expand

$$\alpha(E) \simeq \ell + \left(\frac{d\, \mathrm{Re}\,\alpha}{dE}\right)_{E=E_R} (E-E_R) + i(\mathrm{Im}\alpha)_{E=E_R} \qquad 23.23$$

where $\mathrm{Im}\alpha(E)$ is assumed to be small compared to the real part. Replacing in 23.21

$$f_\ell(E) = \frac{1}{\pi} \frac{\beta(E_R)}{(2\ell+1)\left(\frac{d\mathrm{Re}\,\alpha}{dE}\right)_{E_R}\left(E-E_R+\frac{i\Gamma}{2}\right)} \qquad 23.24$$

where

$$\frac{\Gamma}{2} = \frac{\mathrm{Im}\alpha(E_R)}{\left(\frac{d\mathrm{Re}\,\alpha}{dE}\right)_{E_R}} \qquad 23.25$$

For $E \approx E_B < 0$, $\text{Im}\alpha = 0$, $\frac{\Gamma}{2} = 0$ and 23.24 describes a bound state. For $E > E_{th} > 0$ conversely, $\text{Im}\alpha > 0$ and 23.24 is just a Breit-Wigner resonant amplitude.

If the trajectory $\alpha(E)$ of a Regge pole passes near different values of ℓ (for ℓ real, integer), the same Regge pole can cause the appearance of more than one resonance in different angular momentum states, at different energies. If we plot $\text{Re}\alpha(E)$ versus E, we obtain essentially a Chew-Frautschi plot, if $E \to E^2$, and if these results of potential theory are transferred to the high energy domain. There are still a few details to mention, relative to the potential theory approach to Regge poles. What do we know about the way $\alpha(E)$ moves in the complex ℓ-plane? For this we can determine the derivative

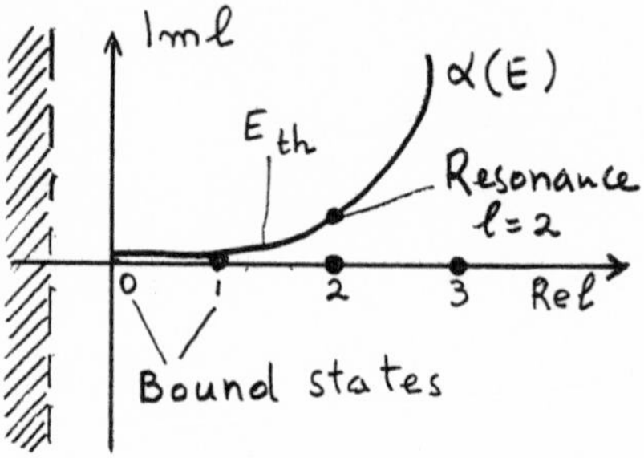

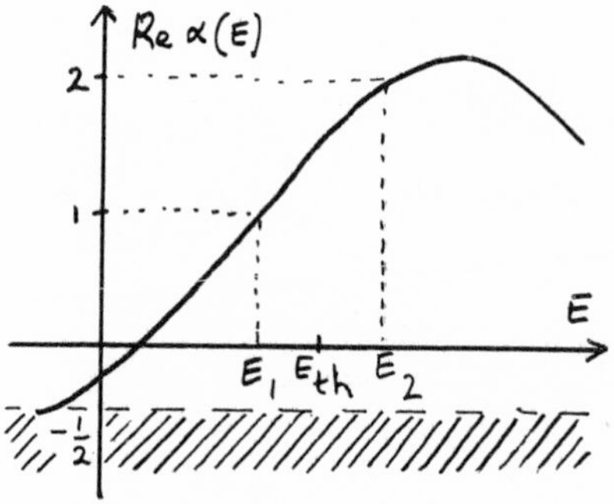

$d\ell/dE$, for $\ell = \alpha(E)$.

Consider a potential well

$$V = \begin{cases} -V & \text{for } r<a \\ 0 & r>a \end{cases} \qquad 23.26$$

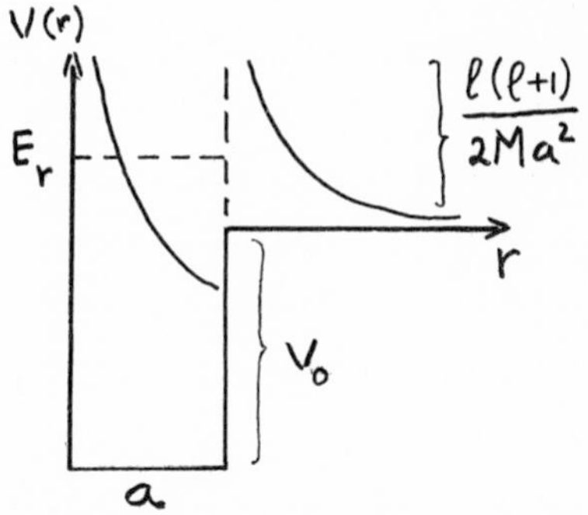

The centrifugal term

$\frac{\ell(\ell+1)}{2Ma^2}$ adds a centrifugal

barrier to the potential.

A resonance at $E = E_R > 0$, for an integer value of ℓ, must be related to the existence of a virtual bound state. Its decay is inhibited by the centrifugal barrier, but decay will eventually occur through tunnelling with a certain probability. As seen above, when $\alpha(E)$ approaches an integer value of ℓ, the scattering amplitude exhibits resonant behaviour. For an integer value of ℓ, Regge poles and virtual bound states are indeed the same thing. Consider two such virtual levels at

$$E_1 \sim \text{const} + \frac{\ell_1(\ell_1+1)}{2Ma^2}$$

$$E_2 \sim \text{const} + \frac{\ell_2(\ell_2+1)}{2Ma^2} \qquad 23.27$$

If these are caused by the motion of the same Regge pole, then we know two points of the Regge trajectory. Extending to non-integer ℓ-values,

$$\frac{d[\ell(\ell+1)]}{dE} = 2Ma^2 \qquad 23.28$$

or, since $E = \frac{p^2}{2M}$,

$$\frac{d|\ell(\ell+1)|}{dp^2} = \frac{d(\ell+1/2)^2}{dp^2} = a^2 \qquad 23.29$$

(valid near real values of ℓ).

This result implies

$$d\ell/dE > 0, \quad \ell > -1/2, \qquad 23.30$$

namely a monotonically increasing trajectory.

How does the $\Delta J = 2$ rule come about? It implies the existence of an exchange potential for the space coordinates in the Schroedinger equation. Since for $\vec{r} \to -\vec{r}$, $\cos\theta \to -\cos\theta$, $P_\ell(\cos\theta) = (-1)^\ell P_\ell(-\cos\theta)$, the usual partial wave expansion contains terms which are even (even ℓ) or odd (odd ℓ) under the symmetry operation. The potential that enters in the radial Schroedinger equation is then of the form

$$V(r) = V_{dir}(r) + (-1)^\ell V_{exch}(r) = \begin{cases} V^+ & \text{for } \ell \text{ even} \\ V^- & \text{for } \ell \text{ odd} \end{cases} \qquad 23.31$$

The potential then is not the same for the even and odd partial waves, and the scattering amplitude will in general be different in the two cases, with different Regge poles. This can be accomodated by writing the amplitude as follows: (See 23.18.)

$$f = f_+ + f_- = \sum_i \frac{\beta_i(E)}{\sin\pi\alpha_i(E)} \left[P_{\alpha_i(E)}(-\cos\theta) + \tau_i P_{\alpha_i(E)}(\cos\theta)\right] + \text{B.I.}$$

23.32

where $\tau_i = \pm 1$ is the signature, or j-parity.

References

The material in this lecture is based on articles by

W. Kummer, Speculations on Experimental Consequences of Regge Poles, CERN Report 62-13(1912).

W. Kummer, Introduction to Regge Poles, Fortschr. Phys. 14 (429), 1966.

See also

B.E.Y. Svensson, High Energy Phenomenology and Regge Poles, Proceedings of the 1967 CERN School of Physics at Rättvik, CERN Report No. 67-24, Vol. II. The latter contains a very extensive bibliography on Regge poles. Also relevant are the other references for Lecture 22.

24

REGGE POLES AND EXCHANGE PROCESSES

We consider once more pion nucleon scattering. Assume that the Regge amplitude is an adequate description of the interaction between hadrons.

The steps to follow are

1) Start with the RSW representation of the amplitude in the <u>t-channel</u> reaction $\pi\pi \to \bar{N}N$.

2) Use crossing symmetry to obtain the s-channel amplitude.

3) Consider the high energy limit in the s-channel to isolate contributions from the Regge poles with the largest real part.

In the first step (neglect the exchange signature for the moment), we rewrite the Regge amplitude 23.18 in the language of the <u>t-channel</u>. (As done previously in the OME model.)

$$\bar{T}(\bar{s},\bar{t}) = \sum_i \frac{\beta_i(\bar{s})}{\sin\pi\alpha_i(\bar{s})} P_{\alpha_i}(\bar{s})(-\cos\theta_t) + \text{B.I.}(\cos\theta_t,\bar{s}) \qquad 24.1$$

(The factor $8\pi\sqrt{s}$ is contained here in $\beta_i(\bar{s})$.)

Using crossing symmetry, we can rewrite at once:

$$T(s,t) = \sum_i \frac{\beta_i(t)}{\sin\pi\alpha_i(t)} P_{\alpha_i}(t)(-\cos\theta_t) + \text{B.I.}(\cos\theta_t,t) \qquad 24.2$$

Remember that

$$\cos\theta_t = \frac{-2(s+\frac{t}{2}-\mu^2-M^2)}{\sqrt{4\mu^2-t}\,\sqrt{4M^2-t}} \qquad \text{(cf. 22.49, p. 240.)} \qquad 24.3$$

The s-channel amplitude is then given by 24.2.

Note that at high energy in the s-channel, $\cos\theta_t \approx -s$. Thus $\cos\theta_t$ becomes numerically very large as s increases. This is no problem since $\cos\theta_t$ is the <u>t-channel</u> scattering angle and 24.2 is the amplitude for the <u>s-channel</u> physical region. By using now the property of the Legendre function 23.19

$$P_{\alpha(t)}(-\cos\theta_t) = k(t)\left(\frac{s}{s_0}\right)^{\alpha(t)} \quad \text{for} \quad s \to \infty \qquad 24.4$$

where constants and s-independent quantities are included in a function $k(t)$. s_0 is an arbitrary scale factor, to obtain a power of a dimensionless number in 24.4. 24.4 then represents the energy dependence of each Regge pole contribution. Since

$$\left(\frac{s}{s_0}\right)^{\alpha} = \left(\frac{s}{s_0}\right)^{\mathrm{Re}\alpha} e^{i\mathrm{Im}\alpha \log\frac{s}{s_0}} \qquad 24.5$$

It follows that the amplitude at high energy is dominated by the Regge pole of highest $\mathrm{Re}\alpha$, the "leading" Regge pole. What about the background integral? From 23.15 we see that $\mathrm{B.I.}(\cos\theta_t, t)$ depends on $\cos\theta_t$ only through $P_\ell(-\cos\theta_t)$. Since the integral is along $\mathrm{Re}\ell = -1/2(+\varepsilon)$, using 23.19 once more,

$$\mathrm{B.I.}(\cos\theta_t, t) \propto s^{-1/2} \quad \text{for} \quad s \to \infty. \qquad 24.6$$

At high s-channel energy, the amplitude then reads:

$$T(s,t) = \Sigma_i \frac{\gamma_i(t)}{\sin\pi\alpha_i(t)} \left(\frac{s}{s_0}\right)^{\alpha_i(t)} + 0(s^{-1/2}) \qquad 24.7$$

where the functions γ_i stand for the products of $\beta_i(t)k(t)$. For several Regge trajectories such that

$Re\alpha_1 \geq Re\alpha_2 \geq Re\alpha_3 \geq -1/2$, large s

$$T(s,t) = \frac{\gamma_i(t)}{\sin\pi\alpha_i(t)} \left(\frac{s}{s_0}\right)^{\alpha_i(t)} + \frac{\gamma_2(t)}{\sin\pi\alpha_2(t)} \left(\frac{s}{s_0}\right)^{\alpha_2(t)} + \ldots \qquad 24.8$$

or a descending series in powers of s.

In order to complete the asymptotic form, we should consider the signature factor. From 23.32 we obtain, by the same approach as above

$$P_{\alpha(t)}(-\cos\theta_t) + \tau P_{\alpha(t)}(\cos\theta_t) \approx k(t) \left| \left(\frac{s}{s_0}\right)^{\alpha(t)} + \tau\left(-\frac{s}{s_0}\right)^{\alpha(t)} \right|$$

$$= k(t)(1+\tau e^{-i\pi\alpha(t)}) \left(\frac{s}{s_0}\right)^{\alpha(t)} \qquad 24.9$$

In 24.9 we have chosen $(-1)^{\alpha(t)} = e^{-i\pi\alpha(t)}$.

This corresponds to having chosen for the analytic continuation a path which approaches the

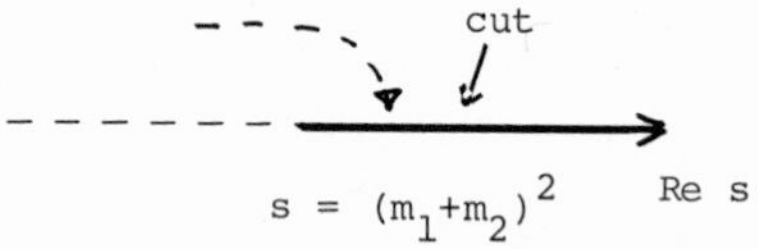

s-positive real axis from above, in order to be consistent with the "+iε" prescription to define the s-channel physical amplitude.

Physical $T(s) = \lim_{\varepsilon\to 0} T(S+i\varepsilon)$, s real.

Alternatively

$$(-s)^{\alpha} = |s|^{\alpha} e^{i\alpha(\theta-\pi)\pi} \underset{\theta\to 0}{\rightarrow} |s|^{\alpha} e^{-i\pi\alpha} , \qquad 24.10$$

In conclusion, the high energy scattering amplitude can be cast in the following form:

$$T(s,t) = \gamma_1(t)\zeta(t)\left(\frac{s}{s_0}\right)^{\alpha_1(t)} + \gamma_2(t)\zeta(t)\left(\frac{s}{s_0}\right)^{\alpha_2(t)} + \ldots \qquad 24.11$$

where

$\alpha(t)$ is the <u>exchanged</u> Regge trajectory

$\gamma(t)$ is the <u>residue function</u>

$\zeta(t)$ is the <u>signature factor</u>, defined as

$$\zeta(t) = \begin{cases} \dfrac{e^{-i\frac{\pi}{2}\alpha(t)}}{\sin\frac{\pi}{2}\alpha(t)} & \text{for } \tau = +1 \\[2ex] \dfrac{i\, e^{-i\frac{\pi}{2}\alpha(t)}}{\cos\frac{\pi}{2}\alpha(t)} & \text{for } \tau = -1 \end{cases} \qquad 24.12$$

To stress once more the concept of t-channel exchange, project the amplitude 24.2, for a single pole, onto a particular partial wave, as done previously. (See 23.24.)

$$T_\ell(s,t) = \frac{1}{\pi} \frac{\beta(t_R)}{(2\ell+1)\left(\frac{d\,\mathrm{Re}\,\alpha}{dt}\right)_{t_R}\left[t-t_R+i\,\frac{\Gamma}{2}\right]} \qquad 24.13$$

where

$$\frac{\Gamma}{2} = \frac{\mathrm{Im}\,\alpha(t_R)}{\left(\frac{d\,\mathrm{Re}\,\alpha}{dt}\right)_{t_R}} \qquad 24.14$$

Note that 24.13 has the form of the Born term in an exchange diagram.

The "exchanged particle" has a mass:

$$m_x^2 = t_R - i\frac{\Gamma}{2}\ ; \quad \text{coupling} \quad g_{aax}(t)g_{bbx}(t) \to \frac{\beta_x(t_R)}{\pi(2\ell+1)\left(\frac{dRe\alpha}{dt}\right)_{t_R}}\ ;$$

$$\text{propagator:} \quad \frac{1}{t - m_x^2}$$

A pictorial summary of what is involved in the exchange of a Regge trajectory in the t-channel, and its effects on the s-channel physical region is as follows:

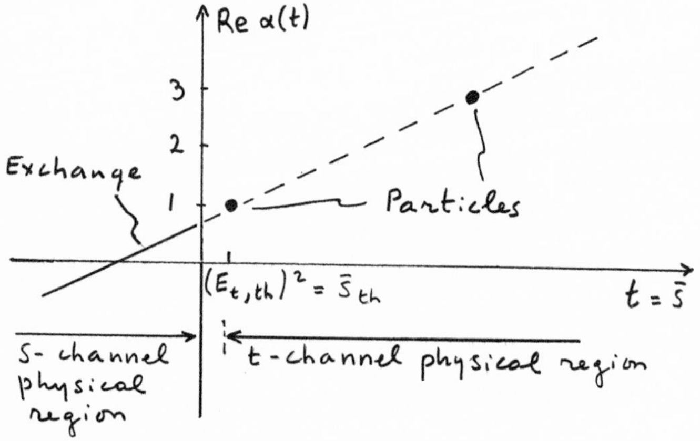

The crucial difference between the OME model discussed previously and the Regge pole exchange model is to be found in the energy dependence of the cross sections.

$$\text{OME} \to \frac{g_a(t)g_b(t)}{t - m^2} P_\ell(\cos\theta_t) \underset{s\to\infty}{\to} F(t)\underset{\substack{\uparrow \\ \text{trouble}}}{s^\ell} \qquad 24.15$$

Going back to the charge-exchange reaction $\pi^- p \to \pi^0 n$, remember that

$$T_{c.e.}(s,t) = \frac{1}{\sqrt{2}}\left[T_{(\pi^+p)} - T_{(\pi^-p)}\right] \qquad 24.16$$

As we have already seen also

$$\frac{d\sigma}{dt} \underset{s\to\infty}{\longrightarrow} \frac{1}{16\pi s^2}|T(s,t)|^2 \qquad 24.17$$

Which are the trajectories which contribute to the charge-exchange reaction? We have to digress first to the total cross sections. From the optical theorem:

$$\sigma_{tot} = \frac{1}{2k\sqrt{s}}\,\mathrm{Im}\,T_{el}(s,0) \underset{s\ \mathrm{large}}{\longrightarrow} \frac{1}{s}\,\mathrm{Im}\,T_{el}(s,t=0) \qquad 24.18$$

Assume that at very high energy only one trajectory (leading trajectory) contributes. Then in order to have $\sigma_{tot} \to$ const., with the energy dependence $s^{\alpha(t)}$ given by 24.11, we must have

$$\alpha_1(t = 0) = 1 \qquad 24.19$$

This is the so-called <u>Pomeranchuk</u> trajectory. To avoid a pole at $t = 0$ (see 24.12, the <u>signature</u> of P must be $\tau = +1$. For $\tau = +1$ and $\alpha_1(0) = 1$, the forward amplitude is <u>purely imaginary</u>. This trajectory assures the validity of the <u>Pomeranchuk theorem</u>:

$$\sigma_{tot}(ab) = \sigma_{tot}(\bar{a}b) \qquad 24.20$$

at asymptotic energies. In fact for

$$ab \to ab \qquad 24.21$$

the t-channel reaction is $a\bar{a} \to \bar{b}b$, for

$$\bar{a}b \to \bar{a}b \qquad 24.22$$

the t-channel reaction is $\bar{a}a \to \bar{b}b$.

The only thing that changes in going from 24.21 to 24.22 is $\cos\theta_t \to -\cos\theta_t$. In order to have equality of σ_{tot}, we need the amplitude to be symmetric in $\cos\theta_t$, therefore $\tau = +1$. ($\tau = -1$ gives an amplitude which is antisymmetric in $\cos\theta_t$, so that $ab \to ab$ and $\bar{a}b \to \bar{a}b$ have amplitudes which are equal in magnitude but opposite in sign.)

The Pomeranchuk pole $\alpha_P(t)$ has the quantum numbers of the vacuum (vacuum trajectory), B=0,S=0,I=0. It can be exchanged in all elastic reactions. At energies which are not asymptotic, the $\pi^{\pm}p$ total cross sections are not the same. One needs at least two more trajectories.

a) A trajectory of negative signature, to contribute with opposite signs to π^+p and π^-p. This is identified with the ρ-trajectory $\alpha_\rho(0) \approx 0.6$.

b) One more vacuum trajectory, P', to explain the energy variation of $\sigma_{tot}(\pi^+p) + \sigma_{tot}(\pi^-p)$, $\alpha_{P'}(0) \approx 0.7$.

In conclusion, need P, P', ρ for elastic scattering.

$$T_{e\ell}(\pi^{\pm}p,s,t) = p + p' \pm \rho \qquad 24.23$$

Going back to 24.16 we see that in $\pi^-p \to \pi^0n$, P and P' cancel out, and

$$T_{c.e.}(s,t) = \sqrt{2}\,\rho \qquad 24.24$$

where, of course, ρ stands for the full Regge amplitude for $\alpha_\rho(t)$.

Now then, from 24.24, 24.17, 24.11 we obtain

$$\left.\frac{d\sigma}{dt}\right)_{c.e.} = D(t)\left(\frac{s}{s_0}\right)^{2\alpha_\rho(t)-2} \qquad 24.25$$

where D(t) includes all the constants, $\gamma_\rho(t)$ and $\zeta_\rho(t)$.

D(t) is unknown from a theoretical standpoint. The essential feature now is that 24.25 has the correct energy dependence, which the OME model did not have. There is more to it. 24.25 predicts the <u>shrinkage of the forward peak</u>. Assume a linear ρ-trajectory

$$\alpha_\rho(t) = \alpha_\rho(0) + \alpha'_\rho t \qquad 24.26$$

then

$$\left(\frac{s}{s_0}\right)^{2\alpha_\rho(t)-2} = \left(\frac{s}{s_0}\right)^{2\alpha_\rho(0)-2} e^{2\alpha'_\rho t \log s/s_0} \qquad 24.27$$

where the exponential factor clearly indicates that the <u>slope</u> of the diffraction peak increases logarithmically with s. In practice, $\alpha_\rho(t)$ is determined experimentally from:

$$\log \left(\frac{d\sigma}{dt}\right)_{c.e.} = \log D(t) + |2\alpha_\rho(t)-2| \log \frac{s}{s_0} \qquad 24.28$$

as a function of log s at fixed t. The result is (Arbab and Chiu, 1966) $\alpha_\rho(t) = (0.56 \pm 0.03) + (0.81 \pm 0.08)t$ 24.29

which extrapolates to a ρ-mass of 740 MeV! (versus 769 MeV for the actual ρ-mass).

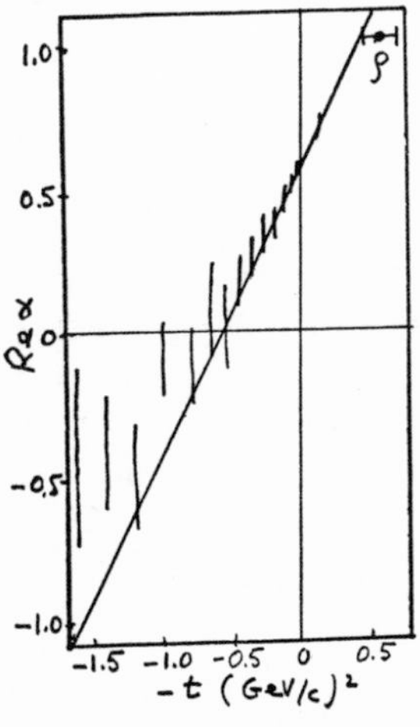

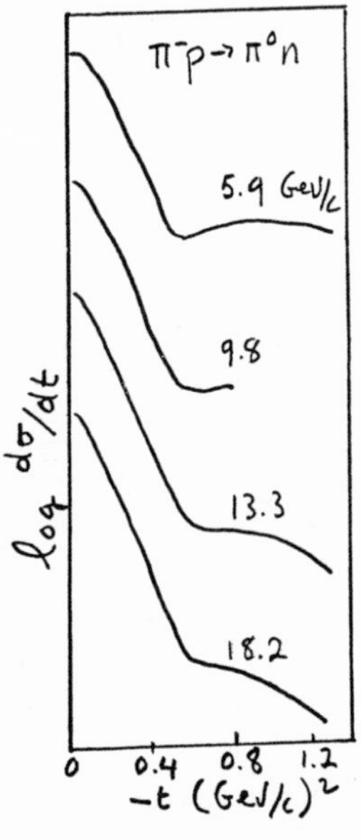

A few remarks at this point. The predicted shrinkage of the diffraction peak is observed in other reactions, like $K^+p \to K^+p$ and $pp \to pp$, but not as clearly in others like $\pi^\pm p$, K^-p and $\bar{p}p$ elastic scattering, at least for energies up to 20-25 GeV/c. The effects of direct channel resonances have a tendency to give an apparent "antishrinkage," in the low and intermediate energy region. In addition, when more than one Regge trajectory contributes to the scattering amplitude, compensating effects may occur.

There are further details worth mentioning, taking as our model reaction still $\pi^-p \to \pi^0 n$, in particular concerning the <u>polarization</u> and the presence of <u>dips</u> in $d\sigma/dt$. Without entering into details, the Regge formalism is extended to describe spin-flip and spin-non-flip amplitudes:

$$8\pi\sqrt{s}\, G\,(\cos\theta, E) = g(t)\zeta(t)\left(\frac{s}{s_0}\right)^{\alpha(t)}, \quad \text{no-flip} \qquad 24.30$$

$$8\pi\sqrt{s}\, H\,(\cos\theta, E) = \sqrt{-t}\; h(t)\zeta(t)\underline{\underline{\alpha(t)}}\left(\frac{s}{s_0}\right)^{\alpha(t)}, \quad \text{flip.} \qquad 24.31$$

↑
important

Note: a) G and H have the same energy dependence.

b) If $\alpha(t)$ is assumed to be <u>real</u>, also the residue functions $g(t)$ and $h(t)$ are <u>real</u>. The phases of G and H are then the <u>same</u>, being given by the same signature factor $\zeta(t)$.

The differential cross section given by 24.30, 24.31 is then:

$$\frac{d\sigma}{dt} = \frac{1}{16\pi}|\zeta(t)|^2\left[|g(t)|^2 + (-t)\alpha^2(t)|h(t)|^2\right]\left(\frac{s}{s_0}\right)^{2\alpha(t)-2} \qquad 24.32$$

and the polarization

$$P = \frac{2\,\mathrm{Im}\,GH^*}{|G|^2+|H|^2} \qquad 24.33$$

It follows that if $\alpha(t)$ = real, P = 0.

This prediction of pure ρ-exchange Regge pole for $\pi^- p \to \pi^o n$ is not verified. In fact a finite polarization of $\sim$ 15% is observed up to $p_{lab} \gtrsim$ 11 GeV/c. The picture then is not as simple. In addition to the ρ-pole exchange, it is necessary to include other singularities, such as Regge cuts, or other complications.

From 24.30, 24.31, however, one gains insight into the cause of the dip observed in $\pi^- p \to \pi^o n$ at $t \sim -0.6$ GeV^2. Fits to the data require h(t) to be comparable in magnitude to g(t). Then, since the spin-flip amplitude contains a factor $\alpha(t)$, it will vanish when $\alpha(t) = 0$. Since the ρ-trajectory crosses $\alpha(t) = 0$ at $t \sim -0.6$ GeV^2, the differential cross sections will have a minimum or dip precisely at t = -0.6 GeV^2, as observed. Thus also the constant-t dip structure of $d\sigma/dt$ is explained. This feature is common to several other elastic processes. What about the polarization in $\pi^\pm p$ scattering? As seen previously, three Regge poles are needed, P, P' and ρ.

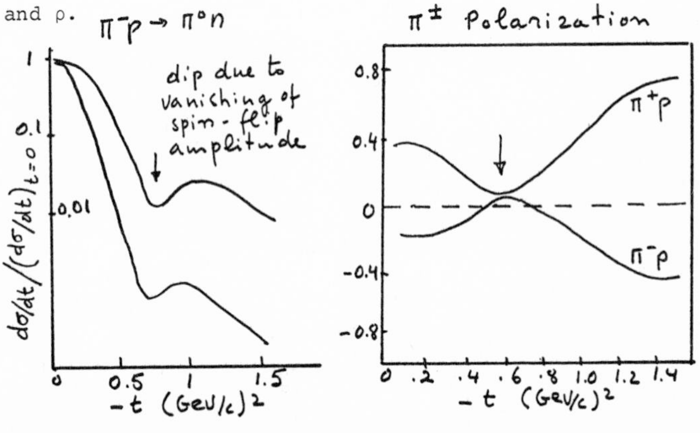

Since data fitting shows that P and P' give very small contributions to the spin-flip amplitude, and conversely ρ has a large spin-flip part, the main contribution to the polarization is given by

$$P_{el} \approx \mathrm{Im}|(G_P+G_{P'})H_\rho^*| \qquad 24.34$$

This has two important consequences:

a) Since the ρ amplitude has opposite signs in π^+p and π^-p, while G_P and $G_{P'}$ are even, the prediction is

$$P_{el}(\pi^+p) = -P_{el}(\pi^-p) \qquad 24.35$$

are observed.

b) When H_ρ vanishes for $\alpha(t) = 0$ at $t \approx -0.6$ GeV/c^2, so does the polarization. Again this is one of the most obvious features of the polarization data.

In conclusion, the sample approach described above indicates some of the achievements of Regge pole phenomenology. This is of course only a very limited and introductory treatment of an increasingly complicated class of phenomena.

References

Much of this lecture follows the approach by B.E.Y. Svensson, High Energy Phenomenology and Regge Poles, Proceedings of the 1967 CERN School of Physics at Rättvik, CERN Report No. 67-24, Vol. II.

Note also the following papers:

F. Arbab and C.B. Chiu, Phys. Rev. 147, 1045(1966).

R.J.E. Sterling, N.E. Booth, G. Conforto, J. Parry, J. Scheid, D. Sherden, A. Yokosawa, Phys. Rev. Letters 21, 1410(1968).

PART II

INTRODUCTION TO HIGH ENERGY PHENOMENOLOGY

B. CONNECTION BETWEEN s- AND t-CHANNEL PHENOMENA

25

GENERALITIES

Duality

In the second part of this course we have touched upon two important descriptions of the interactions of hadrons. On one side we have seen how resonances give a description of the scattering process at low and intermediate energies. We left the subject in fact with the suspicion that resonant states are to be found in essentially all partial waves. On the other side we have analyzed some of the scattering processes at high energy, in terms of the exchange of Regge poles in the crossed channel. Are these two descriptions of the scattering amplitude totally unrelated? Indications at present are that indeed these descriptions are interlocked to the extent that they can be regarded as equivalent descriptions of one and the same property of the interaction of hadrons. This equivalence, in broad terms, is known as duality.

a) Interference models: When we were studying the formation of resonances in the s-channel, we were faced with the problem of describing the amplitude in terms of resonances + "background."

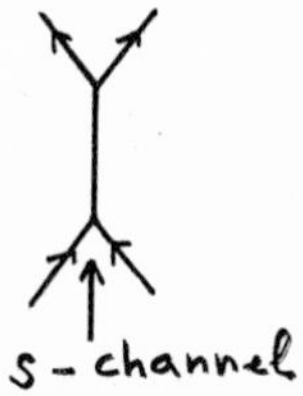

$$A_s = A_{Res} + A_B \qquad 25.1$$

The background term A_B was in general determined as the <u>difference</u> between the sum of the resonant contributions, and the actual amplitude required to fit the data. Similarly, at high energy, we have seen that the amplitude is dominated by <u>exchange</u> processes and we have considered amplitudes of the form

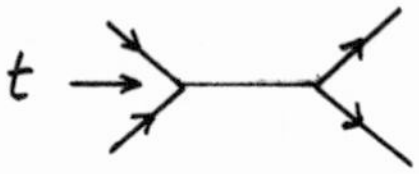

$$A_t = A_{Regge} + B.I. \qquad 25.2$$

Here again, some background, which becomes negligible as $s \to \infty$. For a long time, the relation between A_{Res} and A_{Regge} has been unclear and controversial. It was assumed that the exchange terms were small and could be neglected at low energies, while the opposite was true for the resonances, or that smoothly varying exchange terms were contained in A_B of A_s 25.1. In this case the correct prescription would be, for intermediate energies, to add A_{Regge} and A_{Res}

$$A \approx A_{Regge} + A_{Res} \qquad 25.3$$

This is the prescription of the so-called <u>Regge-Interference Model</u>. On the basis of the so-called "finite energy sum rules," FESR, it has been shown that this approach is indeed <u>wrong</u>. (Dolen, Horn and Schmid, P.R. <u>166</u>, 1768(1968). <u>Duality</u> states instead that

$$A = A_s = A_t \,, \tag{25.4}$$

namely that in principle both A_s and A_t form complete representations of the amplitude, and either one can be used (according to which representation is more convenient in specific energy regions).

b) FESR Bootstrap: The notion implied by 25.4 is actually deeper than apparent. What is suggested in fact is that the baryons (s-channel) may result from the presence of forces associated with meson exchange in the crossed channel and *vice versa*. This would be consistent with the observation that both meson and baryon Regge trajectories have about the same slope of $\sim 1(\text{GeV/c})^{-2}$. The statement of the FESR is that, even at relatively low energies, the *sum* of all resonance contributions, corresponding to a particular set of quantum numbers in the crossed channel, should be well approximated (in a semi-local sense, or in the average over a limited interval) by the *extrapolation* of the Regge amplitude, as determined from the analysis of high energy data. The FESR are conveniently expressed in terms of the variable

$$\nu = \frac{s-u}{2m} \tag{25.5}$$

$$S_n = \int_0^N \nu^n \text{Im}\{F(\nu,t)\}d\nu = \Sigma_i \beta_i(t) \frac{N^{\alpha_i(t)+1+n}}{\alpha_1(t)+1+n} \tag{25.6}$$

The S_n are initially calculated in the low energy part of the physical region of the s-channel ($s > (M_\pi + M_N)^2, t < 0$ in the case of, e.g., π-N scattering). Here the amplitude may be well approximated by a finite number of partial waves

$$F(\nu,t) = \frac{1}{k} \sum_{\ell=1}^{m} (2\ell+1) a_\ell(\nu) P_\ell(\cos\theta(\nu,t) \qquad 25.7$$

(This can be, in particular, a sum of Breit-Wigner resonant amplitudes.) Here the $a_\ell(\nu)$ are taken from the experiment. All the t-dependence occurs through $P_\ell(\cos\theta)$. If the relevant t-channel exchange is already known, e.g., ρ exchange, from the high energy data, then 25.6 provides a consistency condition. The bootstrap aspect arises when 25.6 is evaluated for values of $t > 0$, which belong to the physical region of the crossed channel process (e.g., $\pi\pi \to \bar{N}N$). In this case the amplitude $F(\nu,t)$ on the left-hand side of 25.6 is not accessible to direct measurement. However, the partial wave expansion can be continued into the unphysical region. In the πN case, Dolen, Horn and Schmid have by this procedure obtained a linear trajectory $\alpha_\rho(t)$ consistent with the high energy results, which at $t = M_\rho^2$ has the value $\alpha(M_\rho^2) = 1.0 \pm 0.3$. Thus the location of the t-channel pole is predicted from the knowledge of the s-channel singularities only!

In the following lectures, we shall attempt to make the foregoing qualitative remarks more precise. In particular we shall consider the derivation of dispersion relations and related concepts in order to provide a quantitative discussion of finite energy sum rules.

26

FORWARD DISPERSION RELATIONS

In this and the following two lectures we shall discuss dispersion relations and related topics. We shall go into sufficient detail to provide a firm foundation for reading contemporary literature.

In general dispersion relations (DR's) are based on the assumption ("provable" in limited contexts) that the scattering amplitude is an analytic function of its variables (s,t,u) off the real axis. The best known DR is that for the elastic scattering amplitude in the forward direction. It gives the real part of the amplitude as an integral over essentially the imaginary part or, through the optical theorem, the total cross section. Upon reflection it seems evident that such a relation has little predictive power. Thus to verify such a DR one must measure the forward real part. Of course, if the DR gives the correct results, we tend to believe that the assumptions for its derivation are valid. Indeed, these assumptions are generally used to "derive" other dispersion relations for which mathematical proofs do not exist.

The concept of the finite energy sum rule (FESR) has been greatly discussed in recent years. It follows from essentially the same assumptions as does the forward DR. The great interest in the FESR stems from the fact that it is a potentially predictive concept and that at the very least it seems capable of correlating a great deal of experimental information. The ultimate aim of these lectures is to provide a reasonably precise understanding of this concept.

A Physical Picture

We shall review here some well-known facts concerning dispersion in the classical theory of scattering of electromagnetic waves on various media. By analogy these ideas give us a physical picture of elementary particle scattering.

A medium is said to be dispersive if the index of refraction is a function of the frequency ω of the incident radiation. If at some frequency ω_0 the index of of refraction suddenly drops as shown in the sketch, the dispersion is referred to as anomalous. Physically we can say that the incident radiation is absorbed in the region of ω_0. Consider

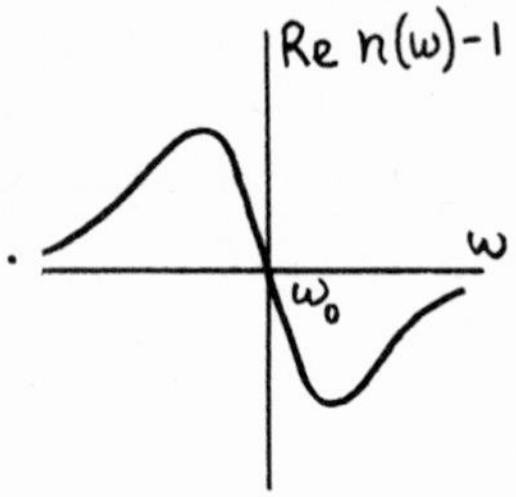

an electric field $E = E_0 e^{i(kz-\omega t)}$ incident upon a harmonically (ω_0) bound electron. If a damping force is present $(\gamma > 0)$ then the motion of the electron is given by

$$m\ddot{x} + m\gamma\dot{x} + m\omega_0^2 x = eE_0 e^{i(kz-i\omega t)}$$

This equation has the steady state solution

$$x = \frac{e/m}{\omega_0^2-\omega^2-i\gamma\omega} E.$$

For N such electrons per unit volume, the electric polarization is

$$\underline{P} = Nex = \frac{K-1}{4\pi} E,$$

from which we find the classical dispersion relation

$$K = 1 + \frac{4\pi Ne^2}{m} \frac{1}{\omega_0^2-\omega^2-i\gamma\omega} .$$

Since $n(\omega) = \sqrt{K}$ we have for reasonable densities

$$n(\omega) \approx 1 + \frac{2\pi Ne^2}{m} \frac{1}{\omega_0^2-\omega^2-i\gamma\omega}$$

or

$$n(\omega)-1 = \frac{2\pi Ne^2}{m} \frac{1}{(\omega_0^2-\omega^2)^2+\gamma^2\omega^2} (\omega_0^2-\omega^2+i\gamma\omega)$$

We may note some interesting analytic properties of $n(\omega)$:

1) $n(-\omega) = n^*(\omega)$ (known as crossing symmetry).

2) For complex ω, $n(\omega)$ has poles at

$$\omega = -i\frac{\gamma}{2}\sqrt{-\gamma^2/4 + \omega_0^2} .$$

Since $\gamma > 0$ for damping, both of these poles lie in the lower half plane, $\text{Im}\omega < 0$.

3) As $|\omega| \to \infty$, we have $|n(\omega)-1| \sim \dfrac{1}{|\omega|^2}$.

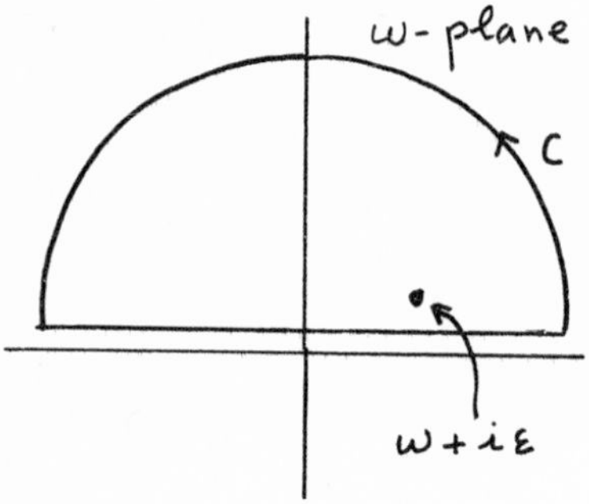

From these properties of $n(\omega)$, we can derive the Kronig-Kramers dispersion relation for $n(\omega)$. Using Cauchy's theorem over the sketched contour, we have from property (2) (i.e., no pole in upper-half ω-plane)

$$n(\omega)-1 = \frac{1}{2\bar{k}i}\oint \frac{n(\omega')-1}{\omega'-\omega-i\varepsilon}\,d\omega'$$

$$= \frac{1}{2\bar{k}i}\int_{-\infty}^{+\infty} \frac{n(\omega')-1}{\omega'-\omega-i\varepsilon}\,d\omega' + \text{Integral over the arc}$$

Property (3) insures that the contribution from the arc vanishes as $|\omega| \to \infty$. We may now use the formal identity

$$\lim_{\varepsilon\to 0+} \frac{1}{\omega'-\omega-i\varepsilon} = \mathcal{P}\frac{1}{\omega'\omega} + i\pi\delta(\omega'-\omega).$$

Thus

$$n(\omega)-1 = \frac{1}{\pi i}\,\mathcal{P}\int_{-\infty}^{\infty} \frac{n(\omega')-1}{\omega'-\omega}\,d\omega',$$

or taking real and imaginary parts

$$\mathrm{Re}\, n(\omega)-1 = \frac{1}{\pi}\int_{-\infty}^{\infty} \frac{\mathrm{Im}\ n(\omega')}{\omega'-\omega}\, d\omega'$$

$$\mathrm{Im}\, n(\omega) = -\frac{1}{\pi}\int_{-\infty}^{\infty} \frac{\mathrm{Re}\ n(\omega')-1}{\omega'-\omega}\, d\omega' \ .$$

Using crossing symmetry, property (1), we find

$$\mathrm{Re}\, n(\omega)-1 = \frac{2}{\pi}\int_{0}^{\infty} \frac{\omega'\,\mathrm{Im}\ n(\omega')\,d\omega'}{\omega'^2-\omega^2} \quad ,$$

the Kronig-Kramers dispersion relation.

We next consider what happens to our incoming wave as it passes through the dispersive medium. Recall that the wave number $k(=2\pi/\lambda)$ is related to the frequency through the dispersion relation

$$k = \frac{n(\omega)\omega}{c} ,$$

where in free space $n = 1$ and $k = k_0 = \frac{\omega}{c}$.

Inside the dispersive medium $n(\omega)$ is complex so that $k = k_r + ik_i$ and we have absorption:

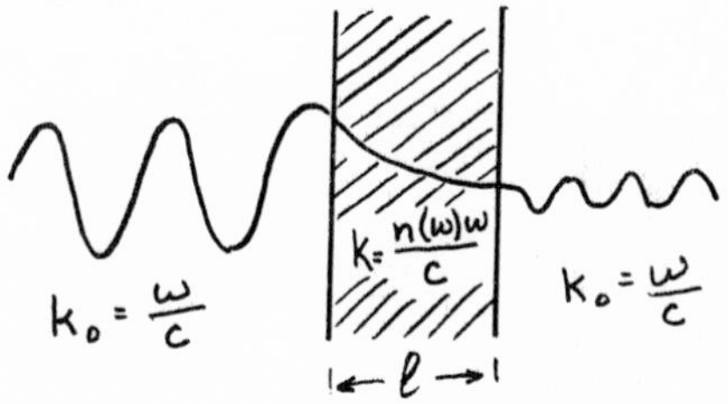

$$E = E_0 e^{ikz-i\omega t} = E_0 e^{ik_r z-i\omega t} e^{-k_i z} \ .$$

For a thin slab of material of thickness ℓ, we find the intensity attenuated by

$$|E|^2 = |E_0|^2 \, e^{-2k_i \ell} \ .$$

This equation is precisely of the form for defining a cross section by transmission; that is, for N scattering centers per unit volume with cross section σ, we have

$$N\sigma = 2k_i = 2\frac{\omega}{c} \text{ Im } n(\omega) \ .$$

Through the optical theorem we know that σ is related to a scattering amplitude $f(\omega)$ in the forward direction:

$$\sigma = \frac{4\pi}{k_0} \text{ Im } f(\omega) = \frac{4\pi c}{\omega} \text{ Im } f(\omega) \ .$$

This suggests that we can relate $n(\omega)$ and $f(\omega)$ through

$$n(\omega) = 1 + \frac{2\pi c^2}{\omega^2} N \, f(\omega) \ .$$

Notice that for the simple model discussed above

$$f(\omega) = \frac{e^2}{m} \frac{\omega^2}{\omega_0^2 - \omega^2 - i\gamma\omega} \ ,$$

a classical Breit-Wigner resonance amplitude. From the integral dispersion relation for $n(\omega)$, it follows that

$$\text{Re } f(\omega) = \frac{2\omega^2}{\pi} \int_0^\infty \frac{\text{Im } f(\omega')d\omega'}{\omega'(\omega'^2 - \omega^2)} \ .$$

This is valid for bound electrons but is incorrect for free electrons where we know that $f(0) = \text{Re } f(0) = -e^2/m$. For free electrons we have

$$\text{Re } f(\omega) = f(0) + \frac{\omega^2}{2\pi^2} \mathcal{P}\int_0^\infty \frac{\sigma(\omega')}{\omega'^2-\omega^2}\, d\omega' \ .$$

How can we apply what we know about classical dispersion to the scattering of elementary particles? First we have an intuitive picture. When a π scatters from a proton, we can think of the π as a plane wave scattering from the dispersive proton medium. Next, while we cannot generally invoke a wave equation to solve $k = n(\omega)\ \omega/c$ in strong interaction, we can hope that the scattering amplitude has sufficient analytic properties such that the integral dispersion relations for $f(\omega)$ still hold. For a beam of massless particles (light), the analytic property that $f(\omega)$ has no poles for $\text{Im } \omega > 0$ follows readily from causality arguments (see Eden and/or Hamilton).

Basic Assumptions

We consider the so-called invariant amplitude T for $t = 0$ (forward direction) normalized so that $\text{Im } T(t=0) = 2p_{cm}\sqrt{s}\ \sigma_{tot} = 2p_{lab} m\ \sigma_{tot}$. We assume that it is an analytic function of z off the real axis, $T(z)$, where

$$\text{Re } z = \nu = \frac{s-m^2-\mu^2}{2m} \ . \qquad 26.1$$

In this expression μ and m are the beam and target masses, s is the cm energy squared, and ν is thus the (total) laboratory energy of the beam particle. We denote by the amplitudes $T_\pm(\nu)$ the scattering process $\pi^\pm p \to \pi^\pm p$ (or $K^\pm p \to K^\pm p$). Furthermore, we assume that $T(z)$ has the following properties:

1) Boundary values and crossing. By this we mean that

$$\left.\begin{aligned} T_-(\nu) &= \lim_{\varepsilon\to 0} T(\nu+i\varepsilon) \\ T_+(\nu) &= \lim_{\varepsilon\to 0} T(-\nu-i\varepsilon) \end{aligned}\right\} , \qquad 26.2$$

where $\varepsilon > 0$. In essence we simply mean here that one function T(z) represents in a definite analytic sense both scattering processes.

2) On the real axis, T(z) has at worst the following properties:

a) cuts extending from ν_- to ∞ and from $-\nu_+$ to $-\infty$.

b) possible (real) poles in the interval $(-\nu_+,\nu)$

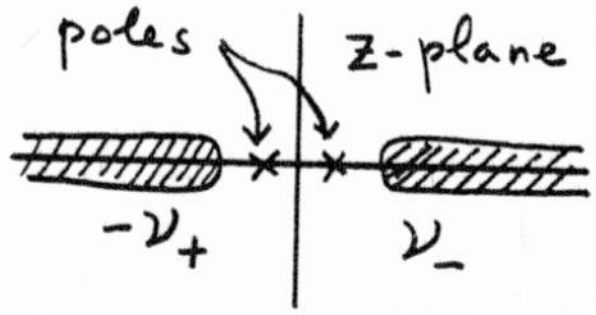

c) T is real and finite in the interval $(-\nu_+,\nu_-)$, except for possible poles.

3) "Polynomial boundedness." For some N, $|T(z)| < z^N$ for all z off the real axis (but on the physical or first sheet).

While we shall not do so here, many of these properties can be proven in the context of certain classes of field theory. Thus, for example, analyticity of T can be proven for restricted regions of t (such as t=0). The essential ingredient is that field operators (anti-)commute for space-like separation; this property of the operators is often referred to as "causality." The ultimate goal of such "proofs" seems to be to reduce the number of axioms needed to describe the strong interaction amplitude to the smallest possible self-consistent set. For our purposes it suffices to accept the above assumptions and to try to understand them from a hopefully more physical point of view.

The crossing property, Equation 2, is perhaps the most difficult one to appreciate physically. This is simply because it is intimately related to the assumption that T(z) is an analytic function off the real axis and on the physical sheet.

Consider the well-known variables s,t,u. Since

$s + t + u = 2\mu^2 + 2m^2$, we

have for $t = 0$

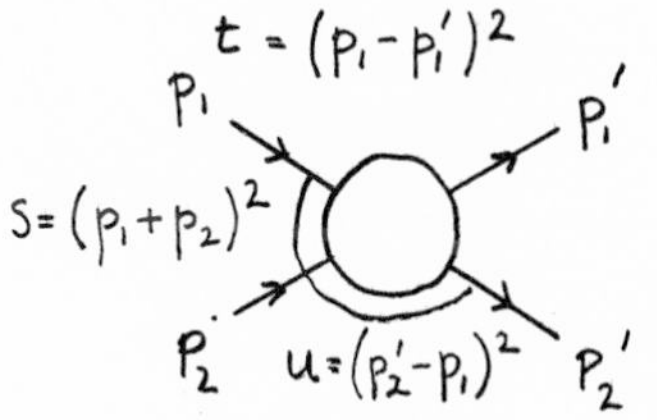

$$\nu = \frac{s-u}{4m} .$$

Hence sending ν into $-\nu$ is equivalent to s going into u. If we consider T_- and T_+ as functions of a complex variable, then Equation 2 implies

$$\left.\begin{aligned} &T_-(\nu+ i\varepsilon) = T_+(-\nu-i\varepsilon) \\ \text{or}\quad & \\ &T_-(s+i\varepsilon,u) = T_+(u-i\varepsilon,s) \end{aligned}\right\} \qquad 26.3$$

This is where the term crossing comes from. (Notice that this relation is often stated without the ε's.) Analyticity is just the assumption that we can analytically continue one complex function to different regions of the s,t,u plane and that the physical amplitudes will be given by the boundary values of this one function. The relation (Equations 2 and 3) between the regions of s,t,u (and thus different physical amplitudes) obtained by analytic continuation is known as crossing symmetry.

The properties of $T(z)$ on the real axis are somewhat more readily appreciated from the notion of unitarity introduced in Lecture 15,

$$\mathrm{Im}T_{ii}(s,t=0) = \frac{1}{2}\sum_j \prod_{i=1}^{j} \left(\frac{\int d^3q_i}{(2\pi)^3 2E_i}\right) |T_{ij}|^2 (2\pi)^4 \delta^4(p_i - p_j) \qquad 26.4$$

where $T_{ii} = T(\nu)$. The crucial points to note are the sum over all physical intermediate states j and the δ^4 function. Below $\nu = \mu$ $(=\nu_-)$ there are in general no physical states available. Thus the delta function says that $T(\nu)$ is real below $\nu = \mu$. As a consequence of this and provided $T(z)$ is analytic off the real axis we have by Schwartz's theorem

$$T(z) = T^*(z^*) \qquad 26.5$$

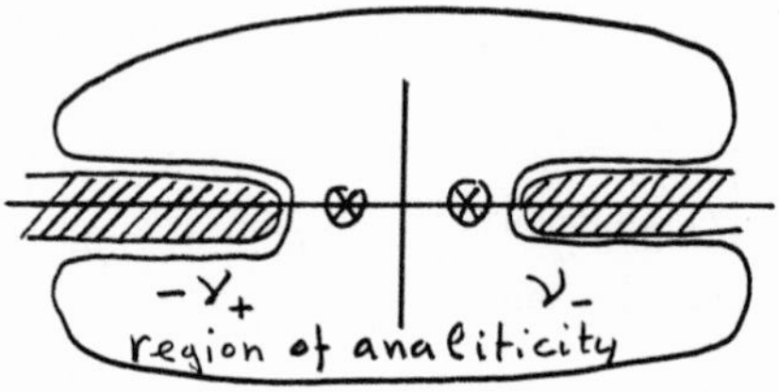

in the region sketched and on the first or physical sheet. We restrict ourselves to the first sheet in these remarks since we know that $T(z)$ has poles on the second sheet which manifest themselves as Breit-Wigner resonances on the real axis.

The existence of poles on the real axis also follows from Equation 4. Thus the nucleon has the same quantum numbers as the πN system; it will contribute to the sum to give a term

$\mathrm{Im}T(\nu) \propto \delta(s-m_N^2)$.

These pole terms, which contribute at isolated points on the real axis, will be discussed in greater detail in the next lecture.

Finally we also observed in Lecture 15 that for two body intermediate states in Equation 4

$$\mathrm{Im}T_{ii}(s,0) \propto \sum_j q_j |T_{ij}|^2 ,$$

where the q_j are the cm-momenta of the (final) two body states. Clearly, each q_j has a cut associated with it starting from the value of ν (or s) where the relevant thresholds open. The first of these is just the elastic threshold $\nu = \mu = \nu_-(=\nu_+)$.

Cauchy's Theorem and Dispersion Relations

We may now apply Cauchy's theorem to our properly defined analytic function $T(z)$. Using the indicated contour and forgetting possible pole terms, we have

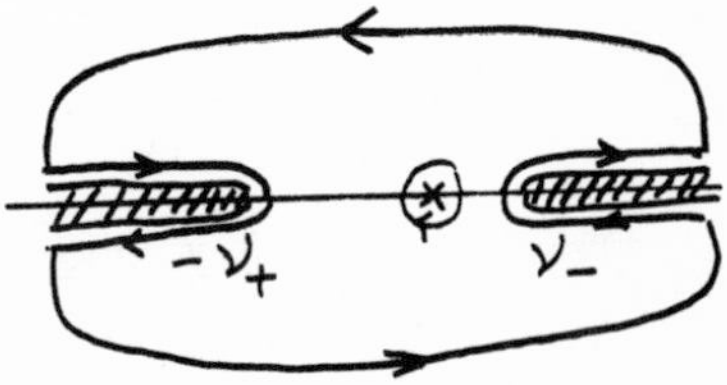

$$T(z) = \frac{1}{2\pi i}\left(\int_{-\infty\pm i\varepsilon}^{-\nu_+\pm i\varepsilon} + \int_{\nu_-\pm i\varepsilon}^{\infty\pm i\varepsilon}\right)\frac{d\nu'}{\nu'-z}\,\Delta T(\nu') + \frac{1}{2\pi i}\int_C \frac{d\nu'}{z'-z}\,T(z').$$

The second term, over the circle at infinity, vanishes as usual provided $|T(z)| < |z|^{-\alpha}$ for $\alpha > 0$. We shall relax this condition later under the assumption of polynomial boundedness.

$\Delta T(\nu)$ is referred to as the "discontinuity across the cut." Clearly we have to take a limit as we approach the real axis from above and below. In other words

$$\Delta T(\nu) = \lim_{\varepsilon\to 0}\left[T(\nu+i\varepsilon)-T(\nu-i\varepsilon)\right] = \lim_{\varepsilon\to 0}\left[T(\nu+i\varepsilon)-T^*(\nu+i\varepsilon)\right]$$

$$= \lim_{\varepsilon\to 0}\left[2i\ \mathrm{Im}T(\nu+i\varepsilon)\right] = 2i\ \mathrm{Im}T(\nu)\ ,$$

where we have used the Schwartz-reflection property of $T(z)$, Equation 5.

Thus,

$$T(z) = \frac{1}{\pi}\left(\int_{-\infty}^{-\nu_+} + \int_{\nu_-}^{\infty}\right)\frac{d\nu'}{\nu'-z}\ \mathrm{Im}T(\nu'). \qquad 26.6$$

To obtain a physical amplitude we shall have to let z approach the real (ν) axis. Setting $z = \nu + i\varepsilon$ and using the identity

$$\frac{1}{\nu'-\nu-i\varepsilon} \underset{\varepsilon\to 0}{=} \mathcal{P}\frac{1}{\nu'-\nu} + i\pi\delta(\nu'-\nu)$$

$$\lim_{\varepsilon\to 0} \mathrm{Re}T(\nu+i\varepsilon) = \frac{1}{\pi}\mathcal{P}\left(\int_{-\infty}^{-\nu_+} + \int_{\nu_-}^{\infty}\right)\frac{d\nu'}{\nu'-\nu}\ \mathrm{Im}T(\nu') \qquad 26.7$$

It is clear that the left hand side of this equation is just $\mathrm{Re}T_-(\nu)$ from crossing, Equation 26.1. We must, however, look more carefully at the previous limiting process for $\mathrm{Im}T(\nu')$. For the integral over the right hand cut, we have

$$\lim_{\varepsilon\to 0} \frac{1}{2i} \int_{\nu_-}^{\infty} \frac{d\nu'}{\nu'-\nu} \left[T(\nu'+i\varepsilon)-T(\nu'-i\varepsilon)\right]$$

$$= \lim_{\varepsilon\to 0} \int_{\nu_-}^{\infty} \frac{d\nu'}{\nu'-\nu} \,\mathrm{Im}T\,(\nu'+i\varepsilon) = \int_{\nu_-}^{\infty} \frac{d\nu'}{\nu'-\nu} \,\mathrm{Im}T_-(\nu') \;.$$

However, the integral over the left hand cut is somewhat different.

$$\int_{-\infty}^{-\nu_+} \frac{\mathrm{Im}T(\nu')}{\nu'-\nu}\, d\nu' = \lim_{\varepsilon\to 0} \frac{1}{2i} \int_{-\infty}^{-\nu_+} \frac{T(\nu'+i\varepsilon)-T(\nu'-i\varepsilon)}{\nu'-\nu}\, d\nu'$$

$$= \lim_{\varepsilon\to 0} \frac{1}{2i} \int_{\infty}^{\nu_+} \frac{T(-\nu'+i\varepsilon)-T(-\nu'-i\varepsilon)}{\nu'+\nu}\, d\nu'$$

$$= \lim_{\varepsilon\to 0} \frac{1}{2i} \int_{\infty}^{\nu_+} \frac{T^*(-\nu-i\varepsilon)-T(-\nu'-i\varepsilon)}{\nu'+\nu}\, d\nu'$$

$$= \lim_{\varepsilon\to 0} \int_{\nu_+}^{\infty} \frac{\mathrm{Im}T(-\nu'-i\varepsilon)}{\nu'+\nu}\, d\nu' = \int_{\nu_+}^{\infty} \frac{\mathrm{Im}T_+(\nu')}{\nu'+\nu}\, d\nu'$$

Combining these results we have

$$\mathrm{Re}T_-(\nu) = \frac{\mathcal{P}}{\pi} \int_{\nu_-}^{\infty} \frac{\mathrm{Im}T_-(\nu')d\nu'}{\nu'-\nu} + \frac{1}{\pi} \int_{\nu_+}^{\infty} \frac{\mathrm{Im}T_+(\nu')}{\nu'+\nu}\, d\nu \;. \qquad 26.8$$

Comment: Equation 26.8 is valid also for any fixed value of t. In this case,

$$\nu = \frac{s-u}{4m} = \frac{s-\mu^2-m^2}{2m} + \frac{t}{4m} = E_{LAB} + \frac{t}{4m} \;.$$

Note that Equation 26.3, crossing symmetry, is still valid.

Subtractions. We have assumed $|T(z)| < z^{-\alpha}$ for $\alpha > 0$ as $|z| \to \infty$. This condition can be relaxed to $|T(z)| < z^{N-\alpha}$ as $|z| \to \infty$ for some N and $0 < \alpha < 1$. We'll illustrate the procedure for $N = 1$.

Consider a fixed point (the "subtraction point") z_1. Then by the assumed analyticity of $T(z)$, we have

$$T(z) - T(z_1) = \frac{1}{2\pi i} \oint dz' T(z') \left(\frac{1}{z'-z} - \frac{1}{z'-z_1}\right)$$

$$= \frac{(z-z_1)}{2\pi i} \oint dt' \frac{T(z')}{(z'-z)(z'-z_1)}$$

$$= \frac{(z-z_1)}{2\pi i} \left(\int_{-\infty}^{-\nu_+} + \int_{\nu_-}^{\infty} \right) \frac{d\nu' \, \Delta T(\nu')}{(\nu'-z)(\nu'-z_1)}$$

$$+ \frac{(z-z_1)}{2\pi i} \int_C \frac{dz' \, T(z')}{(z'-z)(z'-z_1)} ,$$

over the same contour we used above. Again as $|z'| \to \infty$, the integral over the circle vanishes since

$$\left| \frac{T(z')}{(z'-z)(z'-z_1)} \right| < \left| \frac{z'^{(1-\alpha)}}{z'^2} \right| < \left| \frac{1}{z'} \right| .$$

Higher order subtractions are treated similarly. It is clear that polynomial boundedness of the amplitude is essential for deriving dispersion relations.

If we take z_1 on the real axis $(\nu_1 + i\varepsilon)$, then Equation 26.8 becomes with one subtraction

$$\text{Re}T_-(\nu) - \text{Re}T_-(\nu_1) = \frac{\nu-\nu_1}{\pi} \int_{\nu_-}^{\infty} \frac{\text{Im}T_-(\nu')}{(\nu'-\nu)(\nu'-\nu_1)} d\nu'$$

$$- \frac{\nu-\nu_1}{\pi} \int_{\nu_+}^{\infty} \frac{\text{Im}T_+(\nu')d\nu'}{(\nu'+\nu)(\nu'+\nu_1)} , \qquad 26.9$$

where the first integral is taken as principal valued both at ν and ν_1. An often used subtraction point is $\nu_1 = \nu_- = \mu$. Here $\text{Re}T_-(\nu_1)$ is just the scattering length (in the lab system!); notice, too, that as $\nu \rightarrow \mu$, $\text{Im}T_-(\nu) \rightarrow 0$ just as rapidly so that the integral exists at threshold.

REFERENCES

J. Hamilton, Prog. in Nucl. Phys. 8, 143(1960). An introduction to dispersion relations based on several classical physics examples.

H. Pilkuhn, The Interactions of Hadrons, Ch. 6, Wiley & Sons, New York, 1967.

R. J. Eden, High Energy Collisions of Elementary Particles, Cambridge U. Press, 1967.

27
ODDS AND ENDS OF DISPERSION RELATION PHENOMENOLOGY

Pole Terms and Dispersion Relations for $\pi^{\pm}p$ Scattering

We now find the pole term contribution to our dispersion relation for $\pi^- p \to \pi^- p$. From our expression for unitarity, Equation 26.4, keeping only the intermediate neutron state, we have

$$\mathrm{Im}T_-(z) = \frac{1}{2}\sum_{\text{spin}} \frac{d^3p_n}{2m} 2\pi\ \delta^4(q_{\pi^-}+p_p-p_n)\,|T_{\pi^- p\to n}|^2$$

$$= \pi\ \delta(s-m^2)\sum_{\text{spin}}|T_{\pi^- p\to n}|^2 = \frac{\pi}{2m}\delta\left(\nu+\frac{\mu^2}{2m}\right)\sum_{\text{spin}}|T_{\pi^- p\to n}|^2\ , \qquad 27.1$$

where we have used $s = \mu^2+m^2+2m\nu$. To proceed further, we note that the most general form for $T_{\pi^- p\to n}$ is

$$T_{\pi^- p\to n}(s) = \sqrt{2}\ g(s)\bar{u}(n)\gamma_5 u(p)\ , \qquad 27.2$$

where the factor $\sqrt{2}$ is necessary if $g(s)$ is taken to be the neutral pion coupling. The γ_5 means the π's have odd parity (relative to p,n). Using the spinor normalization of Pilkuhn, we may perform the spin sums in Equation 27.1

$$\sum_{\text{spin}}|T_{\pi^- pn}(s)|^2 = -2g^2(s)\bar{u}(p)\gamma_5(\not{p}_n+m)\gamma_5 u(p)$$

$$= -2g^2(s)\bar{u}(p)(-\not{q}_\pi-\not{p}_p+m)u(p) = 2g^2(s)\bar{u}(p)\not{q}_\pi u(p) = 4mg^2\nu\chi_0^T\chi_0$$

in the laboratory frame. (We drop the spinors since we work in the forward direction.) Our final result is

$$\mathrm{Im}T_-(\nu) = 2g^2\pi\nu\delta(\nu+\frac{\mu^2}{2m}) \; . \qquad 27.3$$

The so-called Born-term contribution is thus

$$\mathrm{Re}T_-^{\mathrm{Born}}(\nu) = \frac{1}{\pi}\int_{-\mu}^{\mu} \frac{\mathrm{Im}T_-(\nu^1)}{\nu'-\nu}d\nu' = \frac{2g^2(\frac{\mu^2}{2m})}{\nu+\frac{\mu^2}{2m}} \; . \qquad 27.4$$

Notice that $g^2(s)$ has become $g^2(\nu_B)$, a constant.

We may combine Equation 27.4 with the unsubtracted dispersion relation of the last lecture:

$$\mathrm{Re}T_-(\nu) = \frac{2g^2\nu_B}{\nu+\nu_B} + \frac{\mathcal{P}}{\pi}\int_{\mu}^{\infty} d\nu'\left[\frac{\mathrm{Im}T_-(\nu')}{\nu'-\nu} + \frac{\mathrm{Im}T_+(\nu')}{\nu'+\nu}\right] \; , \qquad 27.5$$

where $\nu_B \equiv \frac{\mu^2}{2m}$ and $\nu_+ = \nu_- = \mu$ (cf. last lecture). The dispersion relation for $T_+(\nu)$ follows immediately from crossing symmetry and Hermitian analyticity. Thus

$$T_+(\nu+i\varepsilon) = T_-(-\nu-i\varepsilon) = T_-^*(-\nu+i\varepsilon) \quad \text{or} \quad T_+(\nu) = T_-^*(-\nu) \; , \qquad 27.6$$

where ν is real. Using $\mathrm{Im}T_\pm(\nu) = 2mk\,\sigma_{tot}^{\pm}(\nu)$, where $k = \sqrt{\nu^2-\mu^2}$ is the laboratory momentum, we have

$$\mathrm{Re}T_\pm(\nu) = \mp\frac{2g^2\nu_B}{\nu\mp\nu_B} + \frac{2m}{\pi}\mathcal{P}\int_{\mu}^{\infty} k'd\nu'\left[\frac{\sigma_{tot}^{\pm}(\nu')}{\nu'-\nu} + \frac{\sigma_{tot}^{\mp}(\nu')}{\nu'+\nu}\right] \qquad 27.7$$

This, unfortunately, is not our final result! The reason is that total cross sections for $\nu \to \infty$ appear to be constant and the integral would diverge.

The situation may be remedied by introducting a subtraction. We do this by writing dispersion relations for the amplitudes

$$M_{\pm}(\nu) = \frac{1}{2m}\frac{1}{(\nu^2-\mu^2)}\left[T_{\pm}(\nu)-\text{Re}T_{\pm}(\mu)(\nu+\mu)\frac{1}{2\mu}+\text{Re}T_{\mp}(\mu)(\nu-\mu)\frac{1}{2\mu}\right] \quad 27.8$$

Before proceeding lets consider how Equation 27.8 was chosen. For large ν' the integral in Equation 27.7 behaves as $\int^{\infty} d\nu'\, \sigma(\infty)$. To make it converge we clearly need at least two powers of ν and thus $\frac{1}{\nu^2-\mu^2}$ in $M_{\pm}$. Next we want $M_{\pm}(\nu)$ to have the same crossing properties as $T_{\pm}(\nu)$-- Equation 27.6 . Notice that $\text{Re}T_{\pm}(\mu)$ are constants (essentially the scattering lengths); thus $M_{+}(\nu) = M_{-}^{*}(-\nu)$ for real ν. Finally notice that as $\nu \to \mu$

$$M_{\pm}(\nu) \underset{\nu\to\mu}{\longrightarrow} \frac{1}{2m}\frac{1}{k^2}\,\text{Im}T_{\pm}(\nu) .$$

However from a scattering length approximation $\text{Im}T_{\pm} \underset{\nu\to\mu}{\longrightarrow} k$, so that $\text{Im}M_{\pm}(\nu) \underset{\nu\to\mu}{\longrightarrow} \frac{1}{k}$. This does not disturb us since $\int_{\mu} \frac{d\nu'}{(\nu'^2-\mu^2)^{1/2}}$ exists.

Finally $M_{\pm}(\nu)$ obeys the same unsubtracted dispersion as $T_{\pm}(\nu)$ and its pole terms follow from those of $T_{\pm}$:

$$\text{Re}M_{\pm}(\nu) = \frac{\mathcal{P}}{\pi}\int_0^{\infty} d\nu'\left[\frac{\text{Im}M_{\pm}(\nu')}{\nu'-\nu} + \frac{\text{Im}M_{\mp}(\nu')}{\nu'+\nu}\right]$$

or

$$\frac{1}{2m(\nu^2-\mu^2)}\left[\mathrm{ReT}_\pm(\nu)-\mathrm{ReT}_\pm(\mu)(\nu+\mu)\frac{1}{2\mu}+\mathrm{ReT}_\mp(\mu)(\nu-\mu)\frac{1}{2\mu}\right]$$

$$=\mp\frac{2g^2\nu_B}{\nu\mp\nu_B}\,\frac{1}{2m(\nu_B^2-\mu^2)}+\frac{\mathcal{P}}{2m}\int_0^\infty\frac{d\nu'}{k'^2}\left[\frac{\mathrm{ImT}_\pm(\nu')}{\nu'-\nu}+\frac{\mathrm{ImT}_\mp(\nu')}{\nu'+\nu}\right].$$

Again using the optical theorem we have finally

$$\mathrm{ReT}_\pm(\nu)=\frac{1}{2}(1+\frac{\nu}{\mu})\,\mathrm{ReT}_\pm(\mu)+\frac{1}{2}(1-\frac{\nu}{\mu})\,\mathrm{ReT}_\mp(\mu)\ \pm\ \frac{2g^2\nu_B}{\nu\mp\nu_B}\,\frac{(\nu^2-\mu^2)}{(\mu^2-\nu_B^2)}$$

$$+\frac{2mk^2}{\pi}\mathcal{P}\int_\mu^\infty\frac{d\nu'}{k'}\left[\frac{\sigma^\pm_{tot}(\nu')}{\nu'-\nu}+\frac{\sigma^\mp_{tot}(\nu')}{\nu'+\nu}\right] \qquad 27.9$$

For reference we note that the usual CM and LAB amplitudes $(\frac{d\sigma}{d\Omega}_{CM}=|f_{CM}|^2)$ are related to T by

$$f_{LAB}=\frac{k}{k_{CM}}\,f_{CM}=\frac{k}{k_{CM}}\,\frac{T}{8\pi\sqrt{s}}=\frac{T}{8\pi m} \qquad 27.10$$

The subtraction constants $\mathrm{ReT}_\pm(\mu)$ are given by

$$\left.\begin{aligned}\mathrm{ReT}_+(\mu)&=8\pi(m+\mu)a_3\\ \mathrm{ReT}_-(\mu)&=8\pi(m+\mu)\frac{1}{3}(2a_1+a_3)\end{aligned}\right\} \qquad 27.11$$

where a_1 and a_3 are the isospin 1/2 and 3/2 scattering lengths (in the CM). Instead of g^2 one often sees $f^2=\frac{g^2}{4\pi}(\frac{\mu}{2m})^2=0.082\pm0.002$. It turns out the a_1, a_3 and f^2 are best estimated by studying the consistency of dispersion relations with the low energy data.

The Scalar Amplitudes A and B

So far our discussion has been restricted to the forward direction, $t = 0$. The reason for this is that crossing only has a simple form for scalar amplitudes. However, for $t \neq 0$ T can be written in terms of scalar amplitudes,

$$T(mm') = \bar{u}_{m'}(p')\left[A(\nu) + \tfrac{1}{2}(\not{q}_\pi + \not{q}_{\pi'})B(\nu)\right]u_m(p), \qquad 27.12$$

for $\pi p \to \pi' p'$ and $\nu = E_{LAB} + t/(4m)$. This equation is the most general Lorentz-invariant matrix element for reactions with one spin 1/2 particle in the initial and final states and for a process which conserves parity. The scalar amplitudes A and B can be related to the spin-nonflip and spin-flip amplitudes g and h. Indeed one finds (Pilkuhn, Ch. 3.9)

$$\left.\begin{aligned} T(1/2,1/2) &= T(-1/2,-1/2) = 8\pi\sqrt{s}\, g \equiv 8\pi\sqrt{s}(f_1 + f_2\cos\theta) \\ T(1/2,-1/2) &= -T(-1/2,1/2) = 8\pi\sqrt{s}\, h \equiv f_2 \sin\theta \end{aligned}\right\}, \qquad 27.13$$

where

$$\left.\begin{aligned} f_1 &= \frac{1}{8\pi\sqrt{s}}\{(E+m)(E'+m)\}^{1/2}\left[A + B(\sqrt{s}-m)\right] \\ f_2 &= \frac{1}{8\pi\sqrt{s}}\{(E-m)(E'-m)\}^{1/2}\left[-A + B(\sqrt{s}+m)\right] \end{aligned}\right\} \qquad 27.14$$

with $E(E')$ being the (cm) energy of the initial (final) proton.

Notice that $1/2(\not{q}_\pi + \not{q}_{\pi'}) = \not{q}_\pi - \not{p}_{p'} + \not{p}_p$. Since $\not{p}_p u(p) = mu(p)$, we have

$$T(m,m') = \bar{u}_{m'}(p')\left[A(\nu) + \not{q}_\pi B(\nu)\right]u_m(p). \qquad 27.15$$

In the forward direction, this becomes

$$T(1/2\ 1/2) = T(-1/2\ -1/2) = 2m[A(\nu)+\nu B(\nu)] \ , \qquad 27.16$$

where $t = 0$ and $\nu = E_{LAB}(\pi)$. From crossing for forward $T(\nu)$, we may "derive" the crossing relations for A and B:

$$T_-(\nu) = 2m[A_-(\nu)+\nu B_-(\nu)] = T_+(-\nu) = 2m[A_+(-\nu)-\nu B_+(-\nu)]$$

or

$$\left.\begin{aligned} A_{\pm}(\nu) &= A_{\mp}(-\nu) \\ B_{\pm}(\nu) &= -B_{\mp}(-\nu) \end{aligned}\right\} \quad \nu \text{ complex} \qquad 27.17$$

We shall assume that Equation 27.17 is valid for all values of t (fixed). Since

$$A_+(\nu+i\varepsilon) = A_-(-\nu-i\varepsilon) = A_-^*(-\nu+i\varepsilon),$$

we find for real ν

$$\left.\begin{aligned} A_{\pm}(\nu) &= A_{\mp}^*(-\nu) \\ B_{\pm}(\nu) &= -B_{\mp}^*(-\nu) \end{aligned}\right\} \quad \nu \text{ real} \qquad 27.18$$

Another set of A,B amplitudes which are important in dispersion relation phenomenology are the so-called crossing even and odd amplitudes

$$\left.\begin{aligned} A^{(\pm)} &= \tfrac{1}{2}(A_+ \pm A_-) \\ B^{(\pm)} &= \tfrac{1}{2}(B_+ \pm B_-) \end{aligned}\right\} . \qquad 27.19$$

From Equation 27.17 it follows that

$$\left.\begin{aligned} A^{(\pm)}(\nu) &= \pm A^{(\pm)}(-\nu) \\ B^{(\pm)}(\nu) &= \mp B^{(\pm)}(-\nu) \end{aligned}\right\} \quad \nu \text{ complex.} \qquad 27.20$$

Before writing dispersion relations for $A_\pm$, we first determine the Born terms. Recall that Equation 27.1 was at a certain point

$$\mathrm{Im}T_-(\nu) = \pi\delta(\nu+\frac{\mu^2}{2m})g^2\ \bar{u}(p)\not{q}_\pi u(p).$$

Comparison with Equation 27.15 suggests

$$\left.\begin{aligned} \mathrm{Im}A_-^{\mathrm{Born}} &= 0 \\ \mathrm{Im}B_-^{\mathrm{Born}} &= \frac{\pi g^2}{m}\delta(\nu+\nu_B) \end{aligned}\right\} \qquad 27.21$$

Until now we have worked with $t = 0$ so $\nu = E_{LAB}$. For $t \neq 0$, we have $\nu \to \nu + t/4m$ and $\delta(E_L+\nu_B) \to \delta(E_L+t/4m+\nu_B-t/4m) = \delta(\nu+\nu_B-t/4m)$. In general then

$$\mathrm{Im}B_\pm^{\mathrm{Born}} = \frac{\pi g^2}{m}\delta(\nu\mp(\nu_B-\frac{t}{4m})), \qquad 27.22$$

where $\nu = (s-u)/(4m) = (s-m^2-\mu^2+t/2)/(2m)$. Thus from $\mathrm{Re}B = \frac{1}{\pi}\int\frac{\mathrm{Im}B}{\nu'-\nu}d\nu'$, we have

$$\left.\begin{aligned} \mathrm{Re}A_\pm^{\mathrm{Born}}(\nu) &= 0 \\ \mathrm{Re}B_\pm^{\mathrm{Born}}(\nu) &= \frac{-g^2}{m(\nu\mp(\nu_B-\frac{t}{4m}))} \end{aligned}\right\}. \qquad 27.23$$

Finally we readily write down dispersion relations for $A_\pm$ and $B_\pm$

$$\mathrm{Re}A_\pm(\nu) = \frac{\mathcal{P}}{\pi}\int_{\nu_0}^{\infty} d\nu' \left[\frac{\mathrm{Im}A_\pm(\nu')}{\nu'-\nu} + \frac{\mathrm{Im}A_\mp(\nu')}{\nu'+\nu}\right] \qquad 27.24$$

$$\mathrm{Re}B_\pm(\nu) = -\frac{g^2}{m(\nu\mp(\nu_B-\frac{t}{4m}))} + \frac{\mathcal{P}}{\pi}\int_{\nu_0}^{\infty} d\nu' \left[\frac{\mathrm{Im}B_\pm(\nu')}{\nu'-\nu} - \frac{\mathrm{Im}B_\mp(\nu')}{\nu'+\nu}\right], \qquad 27.25$$

where $\nu_0 = \mu+t/4m$ and ν (and ν') $= E_{LAB}+t/4m$.

Comments:

1) The minus sign of the second term of the integrand in Equation 27.25 follows from $B_{\pm}(\nu) = -B_{\mp}(-\nu)$ (ν complex).

2) In the forward direction

$$\mathrm{Im}A_{\pm} + \nu \mathrm{Im}B_{\pm} = 2km\sigma^{\pm}_{tot} .$$

This suggests $\mathrm{Im}B_{\pm} \sim \sigma^{\pm}_{tot}$ as $\nu \to \infty$. In that case the integrand in Equation 27.25 goes like $\frac{1}{\nu}(\sigma^{+}_{tot} - \sigma^{-}_{tot})$. If $\sigma^{+}_{tot} = \sigma^{-}_{tot}$ at $\nu = \infty$ and $\sigma^{+}_{tot} - \sigma^{-}_{tot} = \frac{\mathrm{const}}{\nu^{\alpha}}$ with $\alpha > 0$, we see that Equation 27.25 may not need a subtraction.

3) From the above argument it seems a priori evident that Equation 27.24 would need a subtraction. Höhler and Strauss (Z. für Phys. 232, 205(1970)) have noted that no subtraction seems to be needed. Put another way this means

$$\frac{\mathrm{Im}A_{\pm}}{k} \underset{\nu\to\infty}{\to} 0 \quad \text{and} \quad \frac{\nu \,\mathrm{Im}B_{\pm}}{2mk} \underset{\nu\to\infty}{\to} \sigma^{\pm}_{tot} .$$

In the context of Regge-pole theory we associate the so-called Pomeron amplitude with $\sigma^{\pm}_{tot}(\infty)$. These limits suggest that the Pomeron decouples from the A amplitudes.

References

H. Pilkuhn, Interactions of Hadrons, I. Wiley, 1967.

R. G. Moorhouse, Ann. Rev. of Nucl. Sci. 19, 301(1969).

G. Källén, Elementary Particle Physics, Addison Wesley, 1964.

FINITE ENERGY SUM RULES (FESR)

We have now developed our dispersion relation "machinery" to the point where we can study the concept of the FESR in some detail. We shall proceed by first deriving a sum rule for πN scattering, due to Igi and Matsuda (PRL 18, 625),

$$-\pi f^2 + \int_\mu^\infty \left[\mathrm{Im} f^{(-)}(\nu) - \sum_i \beta_i \nu^{\alpha_i} \right] d\nu = 0,$$

where $\mathrm{Im} f^- \propto k(\sigma^-_{tot} - \sigma^+_{tot})$ and the sum runs over all odd signature Regge poles with $\alpha(0) > -1$. When one observes that above a certain energy ν_c the integrand is essentially zero, this sum rule can then be written as a finite energy sum rule,

$$-\pi f^2 + \int_\mu^{\nu_c} \mathrm{Im} f^{(-)}(\nu) d\nu = \sum_i \frac{\beta_i}{\alpha_i + 1} \nu_c^{\alpha_i + 1} .$$

We shall derive a whole class of such FESR's and discuss their more general implications. In particular we will see that in some sense the s-channel resonances are dual or "equal to" the t-channel Regge poles. An important exception to this rule is the Pomeron. As our final topic we shall discuss the hypothesis of Harari and Freund that the Pomeron is dual to the low energy s-channel background.

Preliminaries

To facilitate deriving the Igi-Matsuda (hereafter, I-M) sum rule, we first record several useful relations. Consider the amplitude

$$f^{(-)}(\nu) = \frac{1}{4\pi}\left[A^{(-)} + \nu B^{(-)}\right] = \frac{T^{(-)}}{8\pi m} , \qquad 28.1$$

for $t = 0$. As noted in the last lecture, it is the laboratory amplitude normalized so that $\left.\frac{d\sigma}{d\Omega}\right|_{lab} = |f|^2$.

Using $\mathrm{Im}T(\nu,t=0) = 2k\sigma_{tot}$, we have

$$\mathrm{Im}f^{(-)} = \frac{\mathrm{Im}T^{(-)}}{8\pi m} = \frac{1}{8\pi m}\,\frac{1}{2}\left[\mathrm{Im}T_- - \mathrm{Im}T_+\right] = \frac{k}{8\pi}(\sigma^-_{tot} - \sigma^+_{tot}) , \qquad 28.2$$

where $k = \{\nu^2 - \mu^2\}^{1/2}$. (Note that we are taking $T^{(-)} = \frac{1}{2}(T_- - T_+)$ in contrast to Equation 27.19; there is nothing significant about the sign.)

We readily see that $f^{(-)}$ is odd under crossing using $A^{(-)}(-\nu) = -A^{(-)}(\nu)$ and $B^{(-)}(-\nu) = +B^{(-)}(\nu)$ (see Equation 27.20):

$$f^{(-)}(-\nu) = \frac{1}{4\pi}\left[A^{(-)}(-\nu) - \nu\, B^{(-)}(-\nu)\right]$$

$$= \frac{1}{4\pi}\left[-A^{(-)}(\nu) - \nu B^{(-)}(\nu)\right] = -f^{(-)}(\nu) \qquad 28.3$$

for ν complex. Since

$$f^{(-)}(\nu+i\varepsilon) = -f^{(-)}(-\nu-i\varepsilon) = -f^{(-)*}(-\nu+i\varepsilon),$$

we have $f^{(-)}(\nu) = -f^{(-)*}(-\nu)$ for ν real.

In a similar manner we may also define a crossing even amplitude

$$f^{(+)}(\nu) = \frac{T^{(+)}}{8\pi m} = \frac{1}{4\pi}\left[A^{(+)}(\nu)+\nu B^{(+)}(\nu)\right]. \qquad 28.4$$

In this case we have

$$\mathrm{Im} f^{(+)}(\nu) = \frac{k}{8\pi}(\sigma^{+}_{tot} + \sigma^{-}_{tot}), \qquad 28.5$$

and $$f^{(+)}(-\nu) = f^{(+)}(\nu), \qquad 28.6$$

for ν complex.

To compute the Born term for $f^{(-)}$, recall (t=0) Equation 27.22, $\mathrm{Im}B_- = \frac{\pi g^2}{m}\,\delta(\nu+\nu_B)$; since $\mathrm{Im}B_+(\nu) = +\mathrm{Im}B_-(-\nu)$, we have

$$\mathrm{Im} f^{(-)}_{Born}(\nu) = \frac{\nu}{8\pi}\left[\mathrm{Im}B_-(\mathrm{Born}) - \mathrm{Im}B_+(\mathrm{Born})\right]$$

$$= \frac{g^2}{4\pi}\,\frac{\pi\nu}{2m}\left[\delta(\nu+\nu_B)-\delta(\nu-\nu_B)\right], \qquad 28.7$$

where $\nu_B = \mu^2/(2m)$. Notice again that we take for convenience $B^{(-)} = \frac{1}{2}[B_- - B_+]$.

In what follows we shall refer to $f^{(\pm)}$ in the Regge approximation as $R^{(\pm)}$. Explicitly we write $R^{(\pm)} = \sum_j R_j^{(\pm)}$ where

$$R_j^{(\pm)}(\nu,t) = \beta_j(t)\left(\frac{\nu}{\nu_0}\right)^{\alpha_j(t)} \begin{cases} \dfrac{-1-e^{-i\pi\alpha_j(t)}}{\sin\pi\alpha_j(t)} \\ \\ \dfrac{+1-e^{-i\pi\alpha_j(t)}}{\sin\pi\alpha_j(t)} \end{cases} \qquad 28.8$$

or

$$R_j^{(\pm)}(\nu,t) = \beta_j(t)\left(\frac{\nu}{\nu_0}\right)^{\alpha_i(t)} \begin{cases} i - \cot\dfrac{\pi\alpha_j(t)}{2} \\ i + \tan\dfrac{\pi\alpha_j(t)}{2} \end{cases} .$$

In these expressions β_j and α_j are real for physical t $(t < 0)$, and ν_0 is a scale factor which we shall take to be one. We note that these amplitudes have the correct crossing properties. Thus,

$$R^{(-)}(-\nu) = \beta(-\nu)^{\alpha}\,\frac{+1-e^{-i\pi\alpha}}{\sin\pi\alpha} = \beta\nu^{\alpha}\,e^{i\pi\alpha}\frac{\left[1-e^{-i\pi\alpha}\right]}{\sin\pi\alpha}$$

$$= \beta\nu^{\alpha}\,\frac{e^{i\pi\alpha-1}}{\sin\pi\alpha} = \beta\nu^{\alpha}\left[i - \frac{(1-\cos\pi\alpha)}{\sin\pi\alpha}\right] .$$

But $R^{(-)}(\nu) = \beta\nu^{\alpha}\,\dfrac{1-e^{-i\pi\alpha}}{\sin\pi\alpha} = \beta\nu^{\alpha}\left[i + \dfrac{(1-\cos\pi\alpha)}{\sin\pi\alpha}\right]$ so that $R^{(-)}(-\nu) = -R^{(-)*}(\nu)$ for ν real. Similarly we have $R^{(+)}(-\nu) = +R^{(+)*}(\nu)$.

Comments:

From the point of view of dispersion relations (crossing), the origin of the signature factors is simply to construct amplitudes with the proper crossing properties. In fact I-M use $R^{(-)} \propto [P_\alpha(\nu)-P_\alpha(-\nu)]/\sin\pi\alpha$. To leading order, $P_\alpha(\nu) \sim \nu^\alpha$, this reproduces Equation 28.8. The important point here is that the essential elments of the Regge parametrization are the ν^α energy dependence and the crossing property.

The connection between crossing and signature can be appreciated by recalling (see Equation 22.30) that

$4p_t q_t \cos\theta_t = s - u = 4m\nu.$

An s-channel amplitude that is even or odd under crossing ($\nu \to -\nu$) will only involve t-channel Regge amplitudes of even or odd signature ($\cos\theta_t \to -\cos\theta_t$), respectively.

The Igi-Matsuda (I-M) Sum Rule

Consider the amplitude

$$F^{(-)}(\nu) = f^{(-)}(\nu) - R_\rho^{(-)}(\nu) \, , \qquad 28.9$$

where $R_\rho^{(-)}$ denotes the ρ-Regge amplitude $\alpha_\rho(0) \approx .5)$.

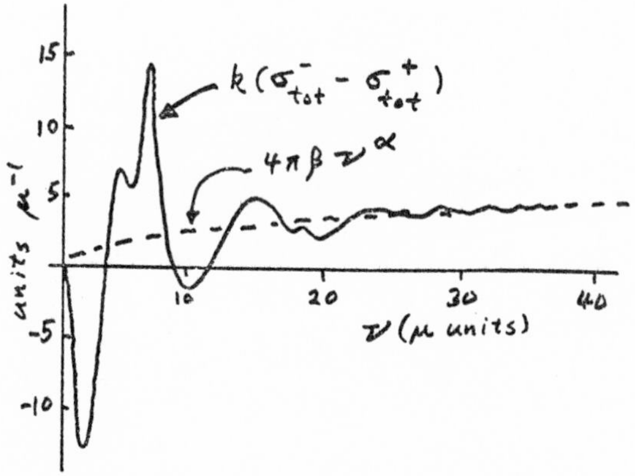

As indicated in the sketch, β_ρ and α_ρ can be chosen so that as $\nu \to \infty$, $F^{(-)}(\nu) \to 0$. In fact let us assume that $\nu F^{(-)} \underset{\nu\to\infty}{\to} 0$; put another way, we assume that the next highest lying trajectory which controls the asymptotic behavior of $f^{(-)}$ has $\alpha(t=0) < -1$.

As in the derivation of Equation 26.6 , we have

$$F^{(-)}(\nu) = \frac{1}{\pi}\left[\int_{-\infty}^{0} \frac{\mathrm{Im}F^{(-)}(\nu')}{\nu'-\nu}\,d\nu' + \int_{0}^{\infty} \frac{\mathrm{Im}F^{(-)}(\nu')}{\nu'-\nu}\,d\nu'\right] .$$

Since $F^{(-)}(\nu)$ is also odd under crossing ($\mathrm{Im}F^{(-)}(\nu) = \mathrm{Im}F^{(-)}(-\nu)$), we have after a change of variables

$$F^{(-)}(\nu) = \frac{2\nu}{\pi}\int_{0}^{\infty} \frac{\mathrm{Im}F^{(-)}(\nu')}{\nu'^2-\nu^2}\,d\nu' \quad .$$

We now apply our assumption about the asymptotic behavior of $f^{(-)}$:

$$\nu F^{(-)}(\nu) = \frac{2}{\pi}\int_{0}^{\infty} \frac{\mathrm{Im}F^{(-)}(\nu')d\nu'}{\left(\frac{\nu'}{\nu}\right)^2 - 1} \underset{\nu\to\infty}{\to} 0 ;$$

that is,

$$\int_0^\infty \mathrm{Im}F^{(-)}(\nu)d\nu = 0 . \qquad 28.10$$

This equation is referred to as a superconvergent sum rule.

The Born term (region of integration 0 to μ) for $F^{(-)}$ is given by that for $f^{(-)}$, Equation 28.7. Thus,

$$-\left(\frac{g^2}{4\pi}\right)\frac{\pi\nu_B}{2m} + \int_\mu^\infty \left[\mathrm{Im}f^{(-)}(\nu)-\mathrm{Im}R_\rho^{(-)}(\nu)\right]d\nu = 0 .$$

Since $\left(\frac{g^2}{4\pi}\right)\frac{\nu_B}{2m} = \frac{g^2}{4\pi}\left(\frac{\mu}{2m}\right)^2 = f^2$ and using Equations 28.2 and 28.8, we have the I-M sum rule,

$$-4\pi f^2 + \frac{1}{2\pi}\int_\mu^\infty \left\{(\nu^2-\mu^2)^{1/2}\left[\sigma^-_{tot}(\nu)-\sigma^+_{tot}(\nu)\right] - 4\pi\beta_\rho\nu^{\alpha_\rho}\right\}d\nu = 0 \qquad 28.11$$

What does this result tell us? First notice that we only have access to experimentally measured cross sections below a finite energy, ν_c. Secondly, we normally think of β_ρ and α_ρ as being determined by the high energy data through fits. Thus as a first test of the validity of Equation 28.11, we can determine β_ρ and α_ρ from the high energy data, assume that $4\pi\beta_\rho \nu^{\alpha_\rho}$ is identical to $k(\sigma^- - \sigma^+)$ for higher energies, and see if Equation 28.11 is satisfied. If it is then our assumptions used to derive Equation 28.11 are consistent with the low energy data.

Suppose, however, that we assume that Equation 28.11 is correct and that above a certain energy $\mathrm{Im}F^{(-)} = 0$. In this case we could predict what β_ρ, say, would be such that the high energy data are well described $(\mathrm{Im}F^{(-)} = 0!)$ by $\beta_\rho \nu^{\alpha_\rho}$. To be specific let's assume $\mathrm{Im}F^{(-)}(\nu) = 0$ for $\nu > \nu_c$; then from Equation 28.10 we have

$$\int_0^{\nu_c} \mathrm{Im}f^{(-)}(\nu)d\nu = \int_0^{\nu_c} \mathrm{Im}R^{(-)}(\nu)d\nu$$

or using Equations 28.2, 28.7, and 28.8,

$$-\pi f^2 + \frac{1}{8\pi}\int_\mu^{\nu_c} \sqrt{\nu^2-\mu^2}\left[\sigma^-_{tot}(\nu)-\sigma^+_{tot}(\nu)\right]d\nu = \frac{\beta_\rho}{\alpha_\rho+1}\nu_c^{\alpha_\rho+1} \qquad 28.12$$

Equation 28.12 is referred to as a finite energy sum rule (FESR). Clearly, if we fix α_ρ, then β_ρ is determined by the low energy data below ν_c.

<u>Comments</u>:

1) If we look at Fig. 1, it is clear that ν_c must be judiciously chosen. Thus if we took $\nu_c \approx 5$ then β_ρ

would be negative! We obtain a better estimate if we take $\nu_c \approx 10\mu$, or, in general, an <u>even</u> number of oscillations.

2) Equation 28.12 strongly suggests--as does Fig. 1--that the high energy Regge amplitude is some sort of average over the low energy data.

Igi and Matsuda found that the approaches to Equation 28.11 just described led to consistent values for β_ρ within a reasonably small range of values for α_ρ (.53-.59). Had such consistency not been obtained, one would have to assume that either one of the usual dispersion relation assumptions (analyticity, crossing) were incorrect or there exist another odd signature Regge trajectory α with $\alpha_\rho(0) > \alpha(0) \geq -1$. In this case Equation 28.12 (and, similarly, Equation 28.11) would become

$$-\pi f^2 + \frac{1}{8\pi}\int_\mu^{\nu_c} (\nu^2-\mu^2)^{1/2}\left[\sigma^-_{tot}(\nu)-\sigma^+_{tot}(\nu)\right]d\nu = \sum_j \frac{\beta_j}{\alpha_j+1}\nu_c^{\alpha_j+1}, \tag{28.13}$$

where j runs over those poles with $\alpha(0) \geq -1$. On the basis of essentially Equation 28.13, I-M in fact argued that certain parameters determined for a postulated ρ' trajectory must in fact be incorrect since they were inconsistent with the available low energy data. The ρ' had been introduced to explain non-zero $\pi^- p \to \pi^0 n$ polarization data (recall that a single ρ trajectory predicts zero polarization); in addition to fitting the data the parameters of any new odd signature Regge pole must also be consistent with Equation 28.13.

Integer Moment FESR's

Consider the amplitudes $\Delta^{\pm}(\nu,t)$, even and odd under crossing, where

$$\Delta^{(\pm)}(\nu,t) = A^{(\pm)}(\nu,t) - R^{(\pm)}(\nu,t) \qquad 28.14$$

and $R^{(\pm)}$ is given by Equation 28.8. In a by now familiar way, we write the fixed t dispersion relation

$$\Delta^{(\pm)}(\nu,t) = \frac{1}{\pi}\int_0^\infty \mathrm{Im}\Delta^{(\pm)}(\nu,t)\left\{\frac{1}{\nu'-\nu} \pm \frac{1}{\nu'+\nu}\right\}d\nu'. \qquad 28.15$$

We assume that for $|\nu| > \nu_c$ $(= E_{LAB} + t/4m)$

$$\left.\begin{aligned} &A^{(\pm)}(\nu,t) = R^{(\pm)}(\nu,t) \\ &\text{or} \quad \Delta^{(\pm)}(\nu,t) = 0 \end{aligned}\right\} \qquad 28.16$$

If we now consider Equation 28.15 for $\nu > \nu_c$ we have by assumption of Equation 28.16

$$0 = \frac{1}{\pi}\int_0^\infty \mathrm{Im}\Delta^{(\pm)}(\nu';t)\left\{\frac{1}{\nu'-\nu} \pm \frac{1}{\nu'+\nu}\right\}d\nu' \ , \quad \nu > \nu_c.$$

Since the integrand, $\mathrm{Im}\Delta^{(\pm)}(\nu'_{,}t)$, vanishes for $\nu > \nu_c$, we can write

$$0 = \frac{1}{\pi}\int_0^{\nu_c} d\nu'\,\mathrm{Im}\Delta^{(\pm)}(\nu',t)\left\{\begin{aligned} &\frac{\nu'}{\nu} + \frac{\nu'}{\nu}^3 + \ldots \\ &1 + \frac{\nu'}{\nu}^2 + \ldots \end{aligned}\right. ,$$

where we have expanded $(\nu'\pm\nu)^{-1}$ in power series. To Satisfy these equations each coefficient of the ν^{-n} must be zero. Using the explicit form, Equation 28.8, from $R^{(\pm)}$. we thus have the so-called moment FESR's

$$S_n^{(+)} = \frac{1}{\nu_c^{n+1}} \int_0^{\nu_c} \nu^n A^{(+)}(\nu,t)d\nu = \sum_j \frac{\beta_j(t)\nu_c^{\alpha_j(t)}}{\alpha_j(t)+n+1} , \qquad 28.17$$

where <u>n is odd</u> (1,2,3,...) and the sum runs over <u>even</u> signature Regge poles (P,P' or f(1260), R or A_2(1300)). Similarly for odd crossing amplitudes, we have

$$S_n^{(-)} = \frac{1}{\nu_c^{n+1}} \int_0^{\nu_c} \nu^n A^{(-)}(\nu,t)d\nu = \sum_j \frac{\beta_j(t)\nu_c^{\alpha_j(t)}}{\alpha_j(t)+n+1} , \qquad 28.18$$

where <u>n is even</u> (0,2,4,...) and the sum runs over <u>odd</u> signature Regge poles (ρ,ρ',ω).

<u>Comments</u>:

1) It must be emphasized that Equations 28.17 and 28.18 apply to any even or odd amplitudes that have the assumed Regge behavior, Equations 28.16 and 28.8. For example $B^{(-)}(\nu,t)$ is even, but is believed to have a $\nu^{\alpha-1}$ energy behavior (see 2) below). Thus for πN scattering (neglecting the ρ'), Equation 28.17 would become

$$S_n^{(+)} = \frac{1}{\nu_c^n} \int_0^{\nu_c} \nu^n \mathrm{Im}B^{(-)}(\nu,t)d\nu = \frac{\beta_\rho(t)\nu_c^{\alpha_\rho(t)}}{\alpha_\rho(t)+n} \qquad 28.19$$

where $n = 1,2,3\cdots$.

2) Instead of $A(\nu,t)$, one often works with

$$A'(\nu,t) = A(\nu,t) + \frac{\nu B(\nu,t)}{(1-t/4m^2)} . \qquad 28.20$$

This is because the t-channel helicity amplitudes are given by

$$T^t_{++;00} = -(t-4m^2)^{1/2} A'$$

$$T^t_{+-;00} = 2\sqrt{t}\, p_t q_t \sin\theta_t\, B/(t-4m^2)^{1/2}.$$

Since it is the partial wave expansion of these amplitudes that is relevant in finding the Regge asymptotic limit, it is more appropriate to express the s-channel helicity (or spin) amplitudes in terms of A' and B or their Regge limits. In addition since (see Pilkuhn)

$$T^t_{+-;00} \propto \sum_J \frac{2J+1}{\{J(J+1)\}^{1/2}} \sin\theta_t\, P_J'(\cos\theta_t) T_J^{++}$$

and $\cos\theta_t \propto \nu$, we have for the Regge poles at $J = \alpha$

$$B \underset{\substack{s\to\infty \\ t\leqslant 0}}{\to} \frac{d}{d\nu} P_\alpha(\nu) \sim \alpha\nu^{\alpha-1} .$$

3) The integrals in Equation 28.17, 28.18, and 28.19 include the Born pole (when B is in the integrand). These are readily found using Equation 28.22 and the convention $B^{(\pm)} = \frac{1}{2}(B_- \pm B_+)$,

$$\mathrm{Im}B^{(\pm)}_{\mathrm{Born}}(\nu,t) = \frac{\pi g^2}{2m}\left[\delta(\nu+\nu_B-\frac{t}{4m}) \pm \delta(\nu-\nu_B+\frac{t}{4m})\right]. \qquad 28.21$$

4) The higher the order of the moment, the better must be the quality of the data (either high or low energy) in order for the FESR's to have predictive power.

Determination of $\beta_\rho(t)$ and $\alpha_\rho(t)$ from Low Energy Data

While the idea of subtracting out the asymptotic behavior of the amplitude is due to Igi, the full implications of the FESR were first elaborated by Dolen, Horn and Schmid (PR 166,1768). We consider here two examples of how the low energy data can determine "high energy" Regge parameters.

Consider the sum rules for $B^{(-)}$,

$$S_n^{(+)} = \frac{1}{\nu_c^n}\int_0^{\nu_c} \nu^n \mathrm{Im}B^{(-)}(\nu,t)d\nu = \frac{\beta_\rho(t)\nu_c^{\alpha_\rho}}{\alpha_\rho(t)+n} ,$$

which were given in Equation 28.19. We can eliminate $\beta_\rho(t)$ from these equations using

$$\alpha_\rho(t) = \frac{3S_3 - S_1}{S_1 - S_3} . \qquad 28.22$$

As indicated in the last lecture Equations 27.13 and 27.14 $B^{(-)}$ can be formed from partial wave amplitudes. Using them as input below $\nu_c = E_{LAB} + t/4m$, DHS found that $\alpha_\rho(t)$ has essentially a linear behavior. In the table their

$t(\mathrm{GeV}^2)$	α_ρ(Eq. 28.22)	α_ρ(HE fit)
m_ρ^2	1.0±0.3	1
0.0	0.4±0.2	0.56-0.7
-0.2	0.3±0.3	0.34-0.40

evaluation of Equation 28.22 is compared with "typical" high energy fitted values of $\alpha_\rho(t)$. Given the quality of the partial wave input and the possibility of a ρ' (cf., DHS) the agreement is satisfactory.

One of the most interesting things shown by DHS was that the FESR could determine the residue of $B^{(-)}$ as a function of t. Recall that this amplitude governs spin or helicity flip in $\pi^- p$ charge exchange (see lecture 24); as noted above and as indicated by the data, we expect the residue to involve $\alpha_\rho(t)$ so that $B_\rho^{(-)}$ has a zero roughly at $t \approx -.5$ or $\alpha_\rho(t) = 0$. We can now appreciate how the low energy data predict this. We again consider Equation 28.19; for the Born contribution, we have using Equation 28.21

$$\nu_c S_1^{(+)} = \int_0^{\mu+t/4m} \frac{\pi g^2}{2m} \nu \left[\delta(\nu+\nu_B \frac{-t}{4m}) - \delta(\nu-\nu_B+\frac{t}{4m})\right] d\nu$$

$$= - \frac{\pi g^2}{2m}(\nu_B - \frac{t}{4m}) = -4\pi^2 \left(\frac{g^2}{4\pi}\right)\left(\frac{\mu}{2m}\right)^2\left(1 - \frac{t}{2\mu^2}\right)$$

$$= -4\pi^2 f^2 \left(1 - \frac{t}{2\mu^2}\right)$$

Thus we get

$$\frac{\beta_\rho(t)\nu_c^{\alpha_\rho(t)+1}}{\alpha_\rho(t)+1} = -4\pi^2 f^2 \left(1 - \frac{t}{2\mu^2}\right) + \int_{\mu+t/4m}^{\nu_c} \nu \mathrm{Im} B^-(\nu,t)d\nu. \qquad 28.23$$

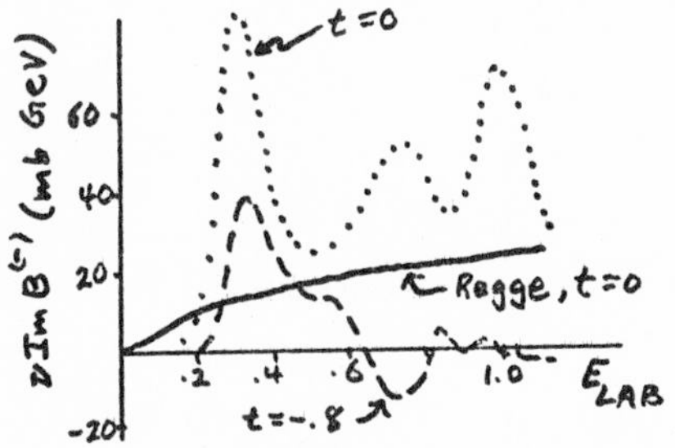

In Fig. 2 we sketch $\mathrm{Im}\nu B^{(-)}$ (after DHS) as obtained from partial wave amplitudes. Notice that as $|t|$ increases, the magnitude of the integral becomes smaller, though positive. However, the Born term we just computed is -1.25 mbGeV^2 at $t = 0$ and -27.6 at $t = -.8$. Since $\mathrm{Im}\nu B^{(-)}$ is surpressed with energy as $|t|$ increases and the Born term becomes more negative, it is clear that the right hand side of Equation 28.23 will change sign. As shown in Fig. 3, Equation 28.23 does in fact indicate that $\beta_\rho(t)$ has a zero between $-.4$ and $-.6$ GeV^2 as expected--here it is predicted through the FESR! For a summary and references to the many applications of the FESR, see J. D. Jackson, Rev. Mod. Phys. 42, 12(1970).

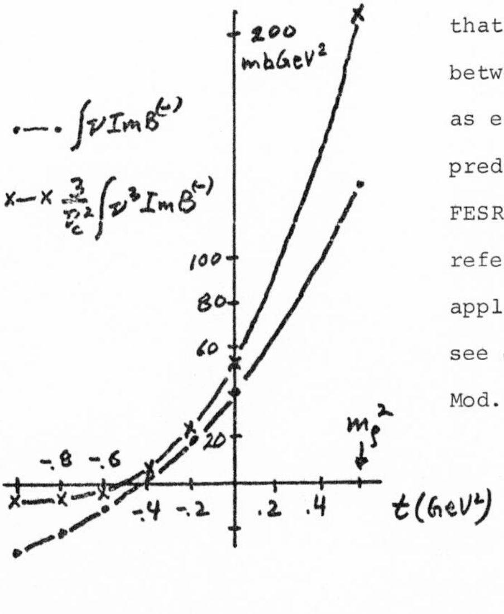

FESR Duality

We have, as yet, not touched upon the interesting feature of Figs. 1 and 2; namely, the amplitudes oscillate about the Regge behavior as a mean. In Fig. 4 is a

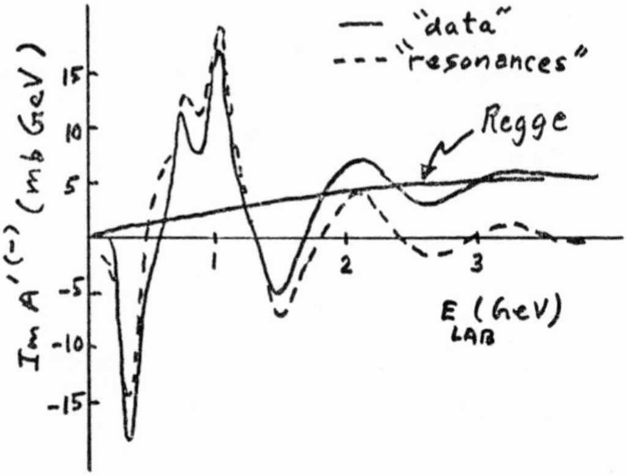

comparison made by DHS. Recall that

$$\mathrm{Im}A'^{(-)}(\nu,0) \propto k[\sigma^-_{tot} - \sigma^+_{tot}] \ .$$

The solid line is interpolated "data" and the dashed line represents what one finds if the amplitude is constructed from the properties of known πN resonances. It would seem that as our knowledge of the higher energy resonances improves, the resonance "model" could readily provide a satisfactory description. In fact DHS point out several examples of where the amplitude can be well explained by adding up the resonances. However, FESR tells us that the Regge amplitude is an average over the low energy amplitude.

In the words of DHS, such observations suggest that "there are two complete representations of any scattering amplitude: One is the partial-wave series which can be dominated by direct-channel resonances or might have a large nonresonating background, and the other is the Regge asymptotic series consisting of pole (and cut) terms s^{α} plus a background integral in the j plane."

Note that Fig. 4 suggests we write the amplitude as ("duality" representation)

$$f = f_{Regge} + f_{Res} - \left\langle f_{Res} \right\rangle ; \qquad 28.24$$

or in words, the total amplitude is the Regge amplitude plus the oscillations of the resonances about their mean. In Equation 28.24 we take

$$\left\langle f_{Res} \right\rangle = \frac{1}{\nu_c} \int^{\nu_c} f_{Res}\, d\nu$$

in the sense of a FESR which would say

$$\left\langle f \right\rangle = \left\langle f_{Regge} \right\rangle . \qquad 28.25$$

While the model of Equation 28.24 would satisfy Equation 28.25, it is clear that the interference model given by

$$f = f_{Regge} + f_{Res} \qquad 28.26$$

will not satisfy Equation 28.25 unless $\left\langle f_{Res} \right\rangle \approx 0$. The fact that in some instances $\left\langle f_{Res} \right\rangle \neq 0$ suggests that Equation 28.26 "double counts"; it can't obey Equation 28.25

because we've somehow put the same information into f twice. It is this vague sense that f_{Regge} and f_{Res} are equivalent which can be referred to as FESR duality. More precise definitions have not been successfully established most likely because there is great freedom in how we "add" Regge and resonance.

The Pomeron and the Harari-Freund Conjecture

The foregoing discussion of FESR duality may be summarized as follows:

1) The experimental amplitude f can be well approximated by simply combining the contributions of known resonances; i.e., f can be "saturated" with resonances.
2) FESR's tell us that $\langle f \rangle = \langle f_{Regge} \rangle$.
3) However, the first observation maintains that $f \approx f_{Res}$. Thus f_{Regge} and f_{Res} are equivalent representations of the amplitude in the sense of the FESR average,

$$\langle f \rangle \approx \langle f_{Res} \rangle = \langle f_{Regge} \rangle . \qquad 28.27$$

Let us now consider precisely for what amplitudes the first point is valid. Until now we have only studied amplitudes which were odd under crossing, $A'^{(-)}$ and $\nu B^{(-)}$. Indeed, the oscillatory behavior (resonances!) of these amplitudes about a mean (Regge!) strongly suggests the validity of Eq. 28.27. (See Figs. 2 and 4). In these instances it is possible, more or less, to "saturate" the amplitudes with resonances.

But what of the crossing even amplitudes? Consider $A'^{(+)}$ for t = 0, recall (Eqs. 28.4 and 28.5) that

$$\mathrm{Im}A'^{(+)}(\nu) = \mathrm{Im}\left[A^{(+)} + \nu B^{(+)}\right] = \frac{k}{2}(\sigma^{+}_{tot} + \sigma^{-}_{tot}). \qquad 28.28$$

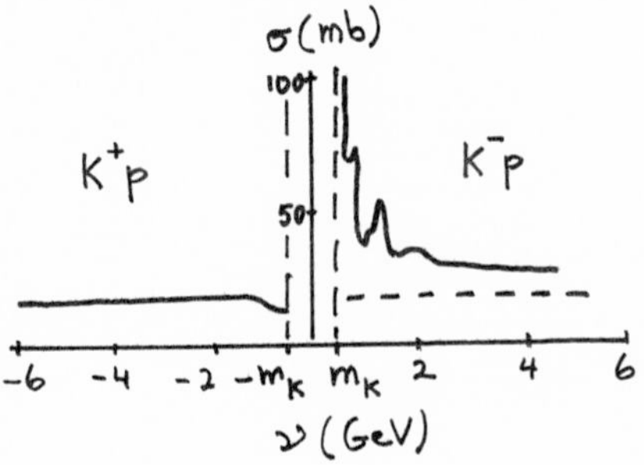

Fig. 5. σ_{tot} for $K^{\pm}p$

Figure 5 (after Jackson) shows, in a crossing symmetric way, the cross sections which enter into Eq. 28.28 in the case of $K^{+}p$ scattering. There are two experimental observations to be made. First in terms of bumps, and as verified by partial wave analyses, there is little or no evidence for resonances in the $K^{+}p$ system. Secondly, if we add the contributions of all known resonances in the $K^{-}p$ system for $\nu \lesssim 1$ GeV, we find we can account for only about half of the observed $K^{-}p$ total cross section. In short, to the best of our present knowledge of the resonance spectrum, it is quite impossible to "saturate" $A'^{(+)}$ with resonances.

From the s-channel viewpoint, that which isn't resonant is, by definition, the background. From the t-channel viewpoint, the distinction between $A'^{(-)}$ and $A'^{(+)}$ (crossing odd and even) is that the Pomeron contributes to $A'^{(+)}$, but not to $A'^{(-)}$, as indicated in Eqs. 28.17 and 28.18. Observations such as these (Freund; Harari) suggest that FESR duality be modified to read

$$\left.\begin{array}{l} \text{for } f^{(-)} \ (\underline{\underline{\text{no}}} \text{ Pomeron}), \quad f^{(-)} \approx f^{(-)}_{Res} \\ \langle f^{(-)}_{Res} \rangle = \langle f^{(-)}_{Regge} \rangle \end{array}\right\} \qquad 28.29$$

28.30

$$\left.\begin{aligned} &\text{for } f^{(+)}\text{(Pomeron)},\ f^{(+)} \approx f^{(+)}_{Backgrd} + (?)\ f^{(+)}_{Res} \\ &\langle f^{(+)}_{Backgrd} \rangle = \langle f^{(+)}_{Pomeron} \rangle \\ &\langle f^{(+)}_{Res} \rangle = \langle f^{(+)}_{Regge} \rangle \end{aligned}\right\} \quad 28.30$$

We see that duality between resonances and (non-Pomeron) Regge poles is maintained. For this to be so, the Pomeron must be "dual" to the low energy background. The ? in the sum in Equation 28.30 is there to emphasize the inherent danger in literally dividing an amplitude into background and resonance contributions. Thus while we can construct <u>models</u> for performing this addition, it should be obvious that what is experimentally accessible is only $f^{(+)}$, the "sum." On the other hand, the qualitative successes of Eqs. 28.29 and 28.30 (see Harari; Harari and Zarmi; Jackson) indicate that they provide a quite reasonable intuitive picture for the connection between s and t channel representations of the scattering amplitude.

References

J.D. Jackson, RMP <u>42</u>, 12(1970).

R.G. Moorhouse, Ann. Rev. of Nucl. Sci. <u>10</u>, 301(1969).

H. Pilkuhn, <u>The Interaction of Hadrons</u>, Wiley, 1967.

K. Igi, PRL <u>9</u>, 76(1962).

K. Igi and S. Matsuda, PRL <u>18</u>, 625(1967).

R. Dolen, D. Horn, and C. Schmid, PRL 19, 402(1967).

H. Harari, PRL <u>20</u>, 1395(1968).

P.G.O. Freund, PRL <u>20</u>, 235(1968).

H. Harari and Y. Zarmi, Phys. Rev. <u>187</u>, 2230(1969).

PROBLEMS

(1) Within the framework of SU(2), explore the transformation properties under an infinitesimal transformation, of the irreducible representation of the product $N \otimes N$ and $N \times \bar{N}$.

(2) Derive explicit expressions for the eigenstates of the irreducible representations of $D(1) \times D(1/2)$ using tensor methods. Explicitly find the irreducible representations of the tensor $\pi^{\beta}_{\alpha} N_{\gamma} = M^{\beta}_{\alpha\gamma}$ and their components.

(3) Assuming SU(2) invariance, find the ratio of the process

$$\frac{K^{-}p \rightarrow \eta\Sigma^{0}}{\bar{K}^{0}p \rightarrow \eta\Sigma^{+}}$$

(4) Given a tensor $M = q \otimes \bar{q}$, show that under an infinitesimal SU(3) transformation

$$e^{i\varepsilon_i \frac{\lambda_i}{2}} \rightarrow 1 + i\varepsilon_i \frac{\lambda_i}{2} ,$$

to first order in ε

$$M' = M + i\varepsilon_i \left[\frac{\lambda_i}{2}, M\right]$$

How will the trace of M transform?

(5) Construct the U^{o}_{1}, U^{0}_{0}, U-spin eigenstates for the pseudoscalar meson octet, making use of the shift operators U_{-} or

U_+ on the appropriate octet members expressed in terms of their quark and antiquark components.

(6) Prove the octet-singlet mixing relation

$$\sin^2\theta = \frac{m_\eta^2 - m_8^2}{m_\eta^2 - m_{\eta'}^2}$$

through reduction of the mass matrix

$$\begin{pmatrix} m_8^2 & m_{18}^2 \\ m_{81}^2 & m_1^2 \end{pmatrix}$$

to diagonal form.

(7) Derive the mass m_1 of the unperturbed SU(3) singlet for the $\frac{3}{2}^-$ baryons, for a mixing angle $\theta = 21.6^\circ$. In other words, we observe $\Lambda(1520)$, how much has its mass been shifted by singlet-octet mixing? Use the mass values given in the notes.

(8) The coupling of the negatively charged baryons $(\Delta^-, \Sigma^-, \Xi^-, \Omega^-)$ belonging to the $J^P = \frac{3}{2}^+$ decuplet to a negative baryon (octet) and a neutral meson (octet) can be described in terms of a single amplitude a. (See 7.20.) Derive the transition matrix elements for

$\Delta^- \to \Sigma^- K^0$	$\Xi^- \to \Sigma^- \bar{K}^0$
$\Sigma^- \to \Xi^- K^0$	$\Xi^- \to \Xi^- \pi^0$
$\Sigma^- \to \Sigma^- \pi^0$	$\Xi^- \to \Xi^- \eta$
$\Sigma^- \to \Sigma^- \eta$	$\Omega^- \to \Xi^- \bar{K}^0$
$\{10\} \to \{8\}\times\{8\}$	$\{10\} \to \{8\}\times\{8\}$

using U-spin considerations.

(9) $\Xi^*(1820)$ has been assigned to the $\frac{3}{2}^-$ baryon octet. Discuss why assignment to a decuplet is ruled out, on the basis of the observed decay modes.

(10) By taking the width of $K^*(892) \to K\pi$ as input, calculate the predicted widths for the decays $\rho \to \pi\pi$ and $\phi \to K\bar{K}$ and compare with experimental widths. Use whatever data you may consider relevant from RPP 70.

(11) Account for the decompositions

$$6 \otimes 6 \otimes 6 = 20 \oplus 56 \oplus 70 \oplus 70'$$

$$6 \otimes 6 = 15 \oplus 21$$

$$35 \otimes 56 = 56 \oplus 70 \oplus 700 \oplus 1134$$

using whatever method you may be able to handle.

(12) Mesons (0^-) travelling along the z-axis collide with a polarized proton target and scatter in the x-z plans. For initial target polarization P_T (along the y-axis) prove that the scattered intensity at angles $+\theta$ and $-\theta$ is given by $I_f(\pm\theta) = I_f(\theta)(1 \pm P_T P_f)$ where $I_f(\theta)\vec{P}_f = 2\text{Im}gh^*\hat{n}$.

(13) Even for the scattering $0^- + \frac{1}{2}^+$, the optical theorem retains its form $\sigma_{tot} = \frac{4\pi}{k} \text{Im } f(0)$. Prove and discuss.

(14) A resonance in the K^-p system is characterized by the following observations:

1) Mass: 1820 MeV ($P_{K^-_{lab}} = 1045$ MeV/c).
2) Width at half-maximum: $\Gamma = 60$ MeV.
3) Elasticity: $x = {}^{\Gamma_{el}}/_{\Gamma} = 0.7$.
4) Size of the resonance, above background, in the elastic channel (K^-p or $\bar{K}^0n$) = 9 mb.
5) The resonance is observed in K^-p, not in K^-n.
6) The angular distribution of elastic scattering requires fourth and fifth order Legendre polynomials.

Determine: a) possible assignments of isospin and spin for the resonant state.

b) A plot of σ_r versus σ_{el} for the energy interval $E_R - 2\Gamma \leq E_R \leq E_R + 2\Gamma$.

c) A plot of the resonant elastic scattering amplitude T_ℓ in the complex plane as a function of energy. (Indicate values of $P_{K^-(lab)}$ on the circle describing the resonance.)

d) A plot of η and δ as a function of energy for the same energy interval as in b).

e) The variation of Γ with energy, for $r_0 = 1$ fermi $= \frac{1}{197}\ \text{MeV}^{-1}$.

f) What is implied in the fact that the fifth order Legendre polynomial is different from zero? (In effect, A_5 for $K^-p \to K^-p$ peaks at 1045 MeV/c for A_5 positive, A_5 for $K^-p \to \bar{K}^0 n$ peaks in the same region but for negative values of A_5. A_4 for both channels shows a positive peak at $\sim$ 1045 MeV/c.)

(15) Following the outline given in eqs. 21.15, 21.16, 21.17 on page 222 of the notes, show that

$$I(\theta_3,\theta_1)=N\{\rho_{33}^{(2)}(\theta_1)+\rho_{11}^{(2)}(\theta_1)+\frac{1}{2}\Big[(\rho_{33}^{(2)}(\theta_1)-\rho_{11}^{(2)}(\theta_1))+ +\sqrt{3}(\rho_{3,-1}^{(2)}(\theta_1)+\rho_{-1,3}^{(2)}(\theta_1))\Big]P_2(\cos\theta_3)\}.$$

(16) Using Eq. 21.24 on page 224 and the following table of C.G. coefficients:

$$(40|\tfrac{5}{2}\,\tfrac{5}{2}\,-\tfrac{3}{2}\,\tfrac{3}{2}) = -\frac{3}{2\sqrt{7}} \qquad (2\,-2|\tfrac{5}{2}\,\tfrac{5}{2}\,-\tfrac{3}{2}\,-\tfrac{1}{2}) = \frac{3}{2\sqrt{7}}$$

$$(40|\tfrac{5}{2}\,\tfrac{5}{2}\,-\tfrac{1}{2}\,\tfrac{1}{2}) = -\frac{1}{\sqrt{7}} \qquad (22|\tfrac{5}{2}\,\tfrac{5}{2}\,\tfrac{1}{2}\,\tfrac{3}{2}) = \frac{3}{2\sqrt{7}}$$

$$(40|\ \tfrac{5}{2}\ \tfrac{5}{2}\ {-\tfrac{5}{2}}\ \tfrac{5}{2}) = -\frac{1}{2\sqrt{7}} \qquad (20|\ \tfrac{5}{2}\ \tfrac{5}{2}\ {-\tfrac{3}{2}}\ \tfrac{3}{2}) = -\frac{1}{2\sqrt{21}}$$

$$(4\ {-2}|\ \tfrac{5}{2}\ \tfrac{5}{2}\ {-\tfrac{3}{2}}\ {-\tfrac{1}{2}}) = \frac{-\sqrt{5}}{2\sqrt{7}} \qquad (20|\ \tfrac{5}{2}\ \tfrac{5}{2}\ {-\tfrac{1}{2}}\ \tfrac{1}{2}) = \frac{2}{\sqrt{21}}$$

$$(42|\ \tfrac{5}{2}\ \tfrac{5}{2}\ \tfrac{1}{2}\ \tfrac{3}{2}) = \frac{\sqrt{5}}{2\sqrt{7}} \qquad (00|\ \tfrac{5}{2}\ \tfrac{5}{2}\ {-\tfrac{3}{2}}\ \tfrac{3}{2}) = \frac{1}{\sqrt{6}}$$

$$(00|\ \tfrac{5}{2}\ \tfrac{5}{2}\ {-\tfrac{1}{2}}\ \tfrac{1}{2}) = -\frac{1}{\sqrt{6}}$$

also eq. (B 90) of Messiah V.I. to find $Y_4^2(\theta_1,0)$, show that,. e.g.,

$$\rho^{(2)}_{3/2,-1/2} = -\frac{\sqrt{2}}{14}\, a(-\tfrac{1}{2})\, a^*(\tfrac{3}{2})\, \{\tfrac{7}{6} - \tfrac{1}{6}P_2(\theta_1) - P_4(\theta_1)\}.$$

(17) Assume a one particle exchange mechanism for the following charge exchange reactions:

a) $K^- p \to \bar{K}^0 n$

b) $np \to pn$

c) $\bar{p}p \to \bar{n}n$

Which particles can be exchanged in each case? (Consider the alternatives allowed by the strong interactions amongst the pseudoscalar mesons π, η, K, the vector mesons ρ, ω, ϕ, K^*, and the 2^+ mesons $A_2(1310)$, $f^0(1250)$, $f^{0'}(1500)$ and $K^*(1420)$.

(18) Let

$$f(k,\theta) = i\frac{k\sigma_\infty}{4\pi}\, e^{b(k)t}$$

where

$$k = \text{c.m. momentum}$$

$$t = -2k^2(1-z),\ z = \cos\theta_{cm}$$

and

$$b(k) = \begin{cases} b_0, \text{ constant} & \text{i)} \\ \alpha' \ln s/s_0 & \text{ii)}; \end{cases}$$

$$\alpha' = \text{constant}, \; s_0 = 1, \; s = E_{cm}^2$$

a) Compute the partial wave amplitudes $T_\ell(k)$. What values may δ_ℓ have? Find η_ℓ.

b) For cases i) and ii) discuss quantitatively the behaviour of T_ℓ and η_ℓ as $k \to \infty$ and as $k \to 0$.

- - - - - - - - -

Useful formulas:

$$\int_{-1}^{1} P_\ell(z) P_{\ell'}(z)\,dz = \frac{2\delta_{\ell\ell'}}{(2\ell+1)}$$

$$\int_{-1}^{1} e^{2zy} P_\ell(z)\,dz = \left(\frac{\pi}{y}\right)^{1/2} I_{\ell+1/2}(2y),$$

where $I_{\ell+1/2}$ is the modified Bessel function of the first kind. Note that

$$I_\nu(x) \underset{x\to 0}{\sim} \frac{(\frac{1}{2}x)^\nu}{\Gamma(\nu+1)} \quad (\nu \neq -1,-2,\ldots)$$

$$I_\nu(x) \underset{x\to\infty}{\sim} \frac{e^x}{\sqrt{2\pi x}} \{1 + 0(\tfrac{1}{x})\}.$$

- - - - - - - - -

(19) Derive the SU(3) relations among the magnetic dipole moments of members of the $J^P = \frac{3}{2}^+$ decuplet.

(20) For the levels of the hydrogen atom, all the eigenstates that have the same number of nodes (zeros) in the radial wave function belong to the same Regge trajectory. What are the energies of the states belonging to the same trajectory? Plot an ℓ versus E diagram for these trajectories.